W0230118

Chemistry: A Cultural Perspective

This book gives a series of stories to illustrate the impact of chemistry on the lives of people from the past to the present day. This book gives chemistry students and lay readers a window into how interactions at the atomic level translate into familiar materials. Teachers of chemistry and related subjects may also find the content and wealth of detail in this book a valuable resource to supplement their lecture room and classroom courses.

Key Features:

- A readable and accessible account of the science underpinning a diverse range of naturally occurring and synthetic substances that are encountered in everyday life.
- Source of engaging material which showcases the multidisciplinary nature of chemistry.
- A text rich in molecular structures, explanatory diagrams and images to promote a thorough grasp of the chemistry presented.
- Dynamic aspect of the chemistry conveyed through a narrative combining historical perspectives and modern developments.
- Inclusion of practical resources shows how concepts can be applied in practice.

Chemistry: A Cultural Perspective

Christine Bladon and Peter Wyatt

CRC Press
Taylor & Francis Group
Boca Raton London New York

CRC Press is an imprint of the
Taylor & Francis Group, an **informa** business

First edition published 2026
by CRC Press
2385 NW Executive Center Drive, Suite 320, Boca Raton FL 33431

and by CRC Press
4 Park Square, Milton Park, Abingdon, Oxon, OX14 4RN

CRC Press is an imprint of Taylor & Francis Group, LLC

© 2026 Taylor & Francis Group, LLC

ISBN: 978-1-032-91273-8 (hbk)
ISBN: 978-1-032-90057-5 (pbk)
ISBN: 978-1-003-56231-3 (ebk)

DOI: 10.1201/9781003562313

Typeset in Times
By codeMantra

Access the support material at www.routledge.com/9781032912738

Contents

Videos of the experiments marked with an asterisk (*) are available in the support material

Preface

This book is about the chemistry behind everyday materials: from medicines to lithium-ion batteries and from metals to artists' pigments. These examples illustrate the contextual nature of the main narrative chapters which are standalone and can be read in any order. This book is aimed at 16- to 18-year-old students and inquisitive adults who are interested in exploring how chemistry is used in real life. Relevant background theory is included in each of these four chapters to equip the reader with the essential knowledge to appreciate the particular areas of chemistry associated with each theme. The focus is very much on applied chemistry, and the material in Chapter 3 ('Riches of the Rocks') is particularly cross-disciplinary and touches on subjects such as geology and archaeology. This book starts with a short introductory chapter and concludes with an experimental chapter which links into content from Chapter 4 ('Chemistry of Colour') and Chapter 5 ('Communication'). This book has developed from the Cultural Perspectives courses that one of us (CMB) delivered at Harris Westminster Sixth Form, London. These courses ran over a single term and were designed to give students the opportunity to study material outside of the A-level chemistry curriculum.

We are appreciative of Queen Mary, University of London granting CMB access to the Chemistry Laboratories in the School of Physical and Chemical Sciences which enabled us to develop the experiments in Chapter 6. We are grateful to colleagues, Ella Taylor at HWSF, John Gorton and our daughter Caroline Wyatt, who have read all or part of the manuscript and have made many valuable suggestions for its improvement.

About the Authors

Christine Bladon has many years of experience teaching chemistry to university and senior students at secondary school. She was Head of Chemistry for seven years at Harris Westminster Sixth Form in London, and her university research interests focused on the bioorganic field. Her publications include *Pharmaceutical Chemistry: Therapeutic Aspects of Biomacromolecules.*

Peter Wyatt is a Senior Lecturer at Queen Mary, University of London. His research in synthetic organic chemistry often involves collaboration with life scientists and physicists. He has contributed more than 70 academic articles to the scientific literature on subjects ranging from amino acids to light-emitting materials.

1 Introduction

Chemistry has been a part of human societies and their cultures for millennia, predominantly in the form of objects that people have made to enrich their lives. Some such objects from former societies have survived to modern times and examining how and when they were made and for what purpose has given current generations insight into the culture of ancient peoples. For example, the reddish-brown pigments in Upper Palaeolithic cave paintings are thought to have been made by finely grinding rocks containing red iron oxide, Fe_2O_3, or by first heating the hydrated iron oxides in yellow goethite and limonite ochres. Heating removes the water, thus forming the red iron oxide according to the equation $2\,FeO(OH) \cdot nH_2O \rightarrow Fe_2O_3 + (2n+1)\,H_2O$. Many of the cave paintings from this period (between 50,000 and 12,000 years ago) depict wild animals and this possibly reflects the importance of animals to the nomadic hunter-gatherer societies at that time.

Historically, indigenous peoples have a strong link to their lands and they have distinct languages, cultures, beliefs and knowledge systems. Some of the oldest living cultures are those of the Australian First Nations peoples and the use of iron oxide-based pigments remains central to several of their cultural practices including body painting and decorating sacred objects. Moreover, these people also developed their own knowledge of the chemical elements and their compounds, for example, calling uranium in its mineral form 'sickness' rocks as they were aware that mishandling them could cause illness.

Some of the smaller stone artefacts excavated in the recent archaeological work at the Ness of Brodgar in Orkney are painted whilst others are engraved with geometric patterns and it may be that people in these Neolithic communities 5,000 years ago just made such items for their aesthetic appeal.[1] More functional items such as arrowheads, axe blades, scrapers and knives were fashioned from rock which could give a sharp cutting edge whilst local sandstones were quarried and shaped into flat blocks for buildings such as at Skara Brae and Maeshowe. The characteristics of different rocks and their potential as raw materials for specific items is what mattered to these communities rather than the inherent chemistry and geology of the individual rock type. Knowledge of this latter aspect came later as chemistry and geology matured into distinct sciences which are connected through the composition of minerals and mineralogy.

The earliest metals used by people were those that occurred in a chemically uncombined form such as gold and copper, and it is likely that the discovery and utilisation of these elements occurred independently in many areas. Sumerians, living around 5,000 years ago in the area which is now southern Iraq, were probably the first people to chemically produce copper through a reduction reaction by heating rocks containing copper minerals in the presence of wood. Copper and tin minerals can sometimes be found in the same geological ore deposit and smelting a mixture of copper and tin ores leading to bronze probably soon followed. Technological innovations which enabled the large-scale production of metals from their ores created a change in the materials available to a society and this metallurgical knowledge helped to

DOI: 10.1201/9781003562313-1

define the culture of the Bronze (2,500–800 BCE) and Iron (800 BCE-43 CE) Ages. The chemical practices which started to develop during these periods together with the early chemical theories which emerged in the following centuries laid the foundations for modern chemistry (Figure 1.1).

The vast majority of the elements in the Periodic Table are metals and a range of chemical processes is now available to extract many of the naturally occurring metals from the ore minerals in rocks. Millions of tonnes of some metals such as iron and aluminium are produced each year, with recycling becoming an increasingly important component in the process. For example, a mobile phone requires aluminium and upwards of 20 other metals in its manufacture including a number of rare earth elements such as terbium, neodymium and dysprosium. Unless recovered through a recycling process, all these elements need to be extracted through mining. There are no mineral deposits for some radioactive metals such as technetium and promethium which have relatively short half-lives compared to the age of the Earth. However, both these elements are part of the uranium decay series and small quantities are formed in the fission products of nuclear fuels.

The geochemistry of rocks and minerals and the significance of mining to past and present communities are explored in Chapter 3. Other sections of Chapter 3 focus on fossils and how rocks and archaeological artefacts are dated. Several archaeological case studies are included in this latter section to afford greater context to the methodology which is based on the radioactive decay of unstable atoms as they transition to a more stable form. Organic botanical materials generally do not survive well in the archaeological record and evidence for the use of plants and plant extracts in the treatment of human disease only started to appear after the advent of writing. Many ancient civilisations started to document accounts of their herbal medicines in the period 2,000–1,000 BCE with information from these remedies becoming more widely distributed after the invention of the printing press in the 15th century CE. By the mid-16th century in the UK, apothecaries, the forerunners of modern pharmacists, were preparing and selling substances for medicinal purposes. Advances in organic chemistry coupled with increasing knowledge surrounding the biological basis of disease led to the development of many effective drugs in the 20th century. Medicinal molecules are the focus of Chapter 2 with progress in the field illustrated with examples taken from the early, mid and later parts of the 20th century. In the final section of the chapter some recent practices and ideas are highlighted to bring the story of molecular medicine into the 21st century.

Colour and communication are the themes of Chapters 4 and 5. Pigments and dyes have been used from ancient times to the present day to bring colour to art work and fabrics and an exploration of the coloured compounds in pigments and dyes is the central component of Chapter 4. The chapter starts by looking into the relation between colour and chemical structure and concludes with a brief overview of the optical interference mechanism which generates iridescent colour. Iridescent colour is a form of visual communication used by some animals and Chapter 5 starts with a section on the pheromones (volatile organic compounds) of some butterflies and moths to illustrate communication by smell. A substantial proportion of Chapter 5 is concerned with the chemistry associated with electronic devices which are transforming human forms of communication and culture. Liquid crystal materials and lithium cobalt oxide ($LiCoO_2$) are integral to the operation of screen technology and

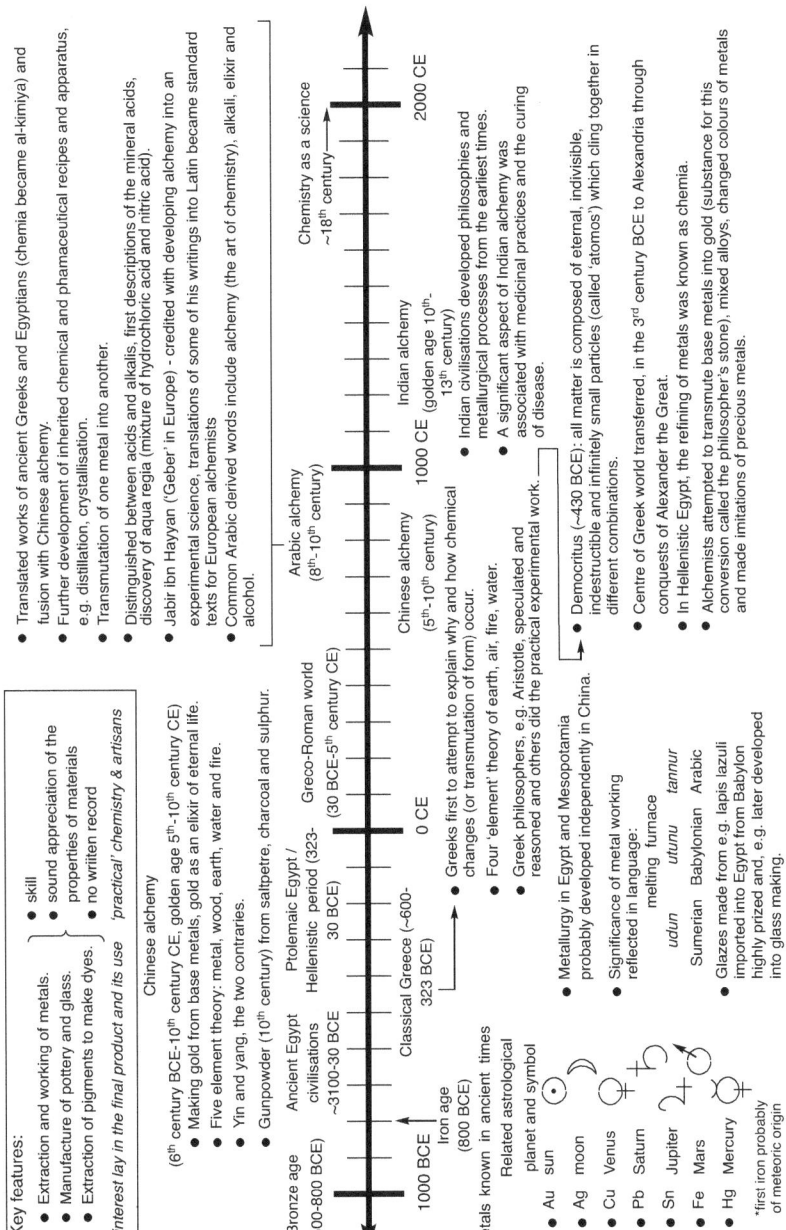

FIGURE 1.1 A timeline of applied chemistry from an artisanal skill to the appearance of modern science. Chemistry can be traced back to early civilisations in Mesopotamia (modern-day southern Iraq), Egypt and China. Early Egyptian and Mesopotamian cultures believed there was a spiritual connection between earthly objects and heavenly bodies and seven metals became associated with the sun, moon and the five planets known at that time. Metallurgists and artisans making pigments, perfumes or glass etc. were skilled workers

(*Continued*)

FIGURE 1.1 (*Continued*)   in their specific area, and as technological expertise developed and spread between different peoples, why and how materials changed during manufacturing processes started to be considered. Greek philosophers speculated in many areas of mathematics and science and devised theories about what occurred during the transformation of one material into another. One theory included the four-element concept in which all visible objects were made from minute indivisible particles of earth, air, fire and water, with change consisting of the separation and/or recombination of these components. Aspects of this theory can be discerned in the modern understanding of some materials, for example, alloys. In making bronze from a molten mixture of copper and tin, some of the copper atoms are substituted with those of tin. Aristotle attached two of the four qualities: hot, cold, moist and dry to each element, the proportions of which varied during change. For example, water was moist and cold, and by overcoming the cold quality by adding heat, water was converted into air which was moist and hot. Aristotle's views however dominated scientific thought and tended to drive out competing ideas such as the atomistic theories of Democritus and these did not reappear until the Renaissance. Philosophers used reasoning and logic for their explanations but did not perform practical work as this was viewed as work for the hands not the mind. During the Hellenistic period, philosophers started to perform experiments to support their ideas and this developed into alchemy which took off in the early centuries of the Common Era. Alchemy was practised independently in several different cultures and aimed to create an elixir of youth, consumption of which would confer immortality, and to discover how base metals such as lead could be converted into gold. This latter process was considered to require an unknown substance, later becoming known as the philosopher's stone, which was also sought by the alchemists. Notable periods in Chinese and Arabic alchemy were the 5th–10th and 8th–10th centuries respectively, with the work of Jabir ibn Hayyan being particularly influential in the Islamic world. Indian alchemy flourished in the 10th–13th centuries and was more concerned with medicine and mineral remedies for specific diseases than transmutation of base metals into gold. Although alchemists did not succeed in their aims, their work greatly increased the knowledge of elements and compounds; they invented many experimental techniques and quantitative methods came to be accepted as essential to chemical investigations. The combination of these factors made possible the foundation of modern chemistry by Lavoisier at the end of the 18th century.

battery power respectively and the chemistry of these compounds is considered in Sections 5.3 and 5.4. Digital photography and photo messaging have changed the way images are taken and shared in the 21st century and the parallel phenomenon in the 19th century was silver halide photography. Silver halide photography could be considered to be chemical magic and Chapter 5 concludes with the chemistry of taking, developing and printing traditional black-and-white photographs.

Chemistry can broadly be split into two branches: 'pure' which focuses on fundamental principles and 'applied' in which these principles are used to develop, often in a technological manner, a solution to a real world situation. The chemistry topics described for the main content of each chapter fall largely into the latter category, but elements of pure chemistry are included to give a more rounded sense to the applied material. Chemistry is also a practical subject as experiments and observations test new theories, or new theories may develop as a consequence of findings from laboratory work. The final chapter gives experimental procedures which enable readers themselves to explore in the laboratory some of the material covered in Chapters 4 and 5. Suggestions for practical work associated with rocks, minerals and fossils are integrated into Chapter 3.

NOTE

1 The Ness of Brodgar site is on the west side of the main island of the Orkney archipelago which is situated off the north coast of Scotland. The Ness of Brodgar is at the centre of the Neolithic landscape in Orkney which is a World Heritage Site.

FURTHER READING

Brock, W. H. (2016) *The History of Chemistry: A Very Short Introduction*. Oxford: Oxford University Press.

Emsley, J. (1998) *The Elements*. 3rd Ed. Oxford: Clarendon Press.

Rohrig, B. (2015) Smartphones: Smart Chemistry. *ChemMatters*. [Online] April/May. pp. 10–12. Available from: https://www.acs.org/content/dam/acsorg/education/resources/highschool/chemmatters/documents/smartphones-smart-chemistry.pdf. [Accessed: 20th June 2025].

The Ness of Brodgar (2024) *The Ness of Brodgar Project – Investigating a Prehistoric Complex in the Heart of Neolithic Orkney*. [Online] Available from: https://www.nessofbrodgar.co.uk. [Accessed: 9th December 2024].

2 Medicinal Molecules

2.1 INTRODUCTION

Medicinal molecules are small organic compounds that are used to prevent or cure diseases in humans and animals. These compounds exert their effect by interacting with biological structures such as receptors or enzymes, which in turn triggers a series of steps which ultimately results in a macroscopic physiological change which benefits the individual.

Historically, our ancestors used preparations of natural products for their medicinal purposes. These folklore remedies were often mixtures of plant products, with herbs forming the bulk of the preparations, and were found through trial and error. As chemical techniques developed, the active constituents were isolated from the crude plant source, purified, structurally characterised and, in due course, many were synthesised in the laboratory. Some isolated compounds proved to be satisfactory as therapeutic agents but others proved to be unsatisfactory, such as being too toxic or by giving unwanted side effects through interfering with biological processes other than that intended.

The search to find less toxic medicinal molecules resulted in the introduction of synthetic compounds as therapeutics in the late 19th century and these became widespread in the 20th century. Synthesising and testing a series of structurally related compounds based on an initial lead molecule became a central feature of the classical medicinal chemical methodology of the mid-20th century.

The emergence of genetic sciences and the groundbreaking work on protein and DNA structure and synthesis in the mid-20th century had an enormous impact on the understanding of the biochemical mechanisms of disease processes. The rational design of new therapeutics using knowledge of the molecular identity of the biological targets thus became a realistic strategy. The advent of tools for genetic engineering such as recombinant DNA (rDNA) methodology and the development of monoclonal antibody techniques in the latter part of the 20th century, together with the synergistic growth in biotechnology and computational techniques, have made drug discovery in the 21st century into a truly interdisciplinary endeavour.

The term 'drug' is often a synonym for medicines or medicinal molecules but can have negative connotations when used to describe the use of chemical substances for improper or recreational purposes rather than therapeutic use. However, the word 'drug' will be used in a positive sense in this chapter.

A number of well-known medicinal compounds with their common name, type of disease targeted and date of introduction as therapeutics are shown in Figure 2.1 along with a selection of scientific and general events of the time to give some additional contextual information.

Five medicinal molecules or families of compounds: aspirin, opioid analgesics, β-blockers, angiotensin-converting enzyme (ACE) inhibitors and insulin, have been chosen to illustrate various aspects of the drug discovery process during the 20th century. These accounts focus on the role of chemistry and the structures of compounds in the design and development phase of these well-known therapeutics.

DOI: 10.1201/9781003562313-2

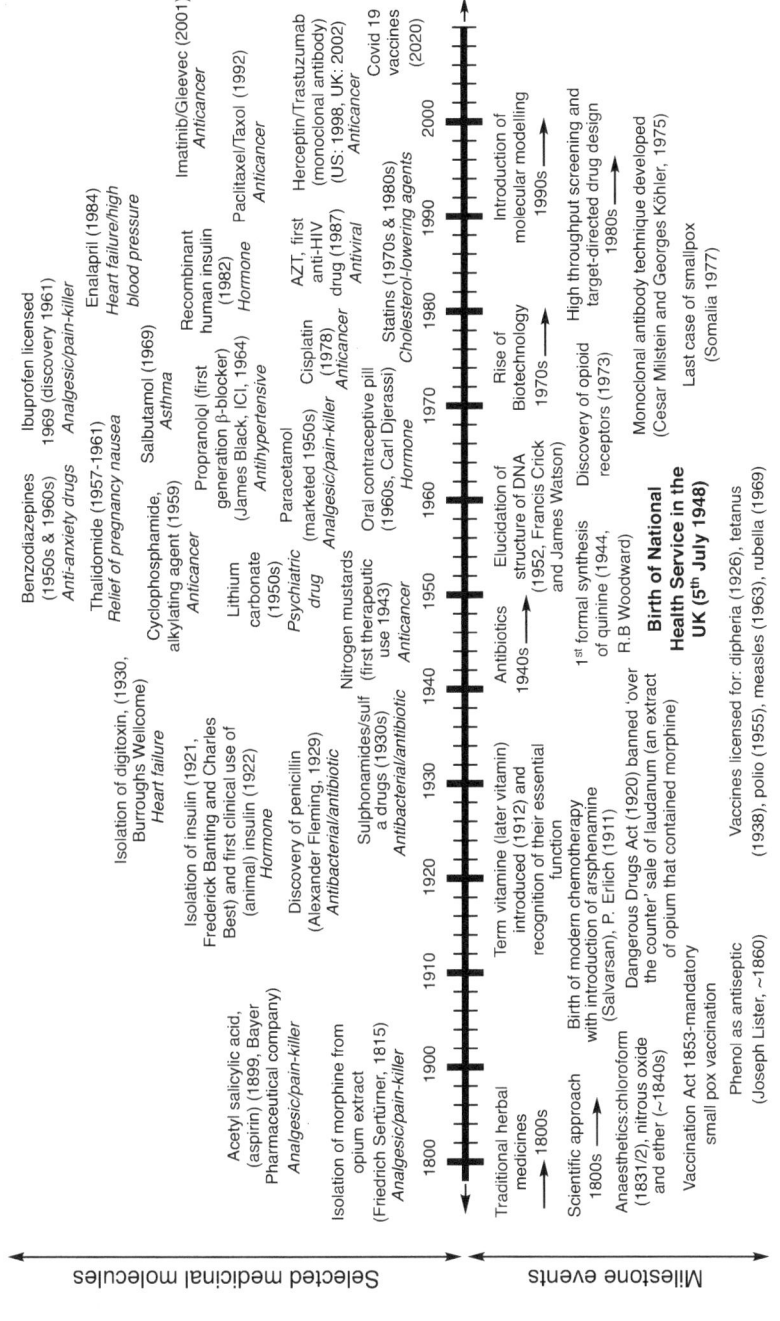

FIGURE 2.1 (a) A selection of familiar drugs with an indication of the type of compound, condition or disease targeted and the date when they were introduced into clinical use. (b) Structures for the small molecule medicinal compounds given in Figure 2.1a.

(Continued)

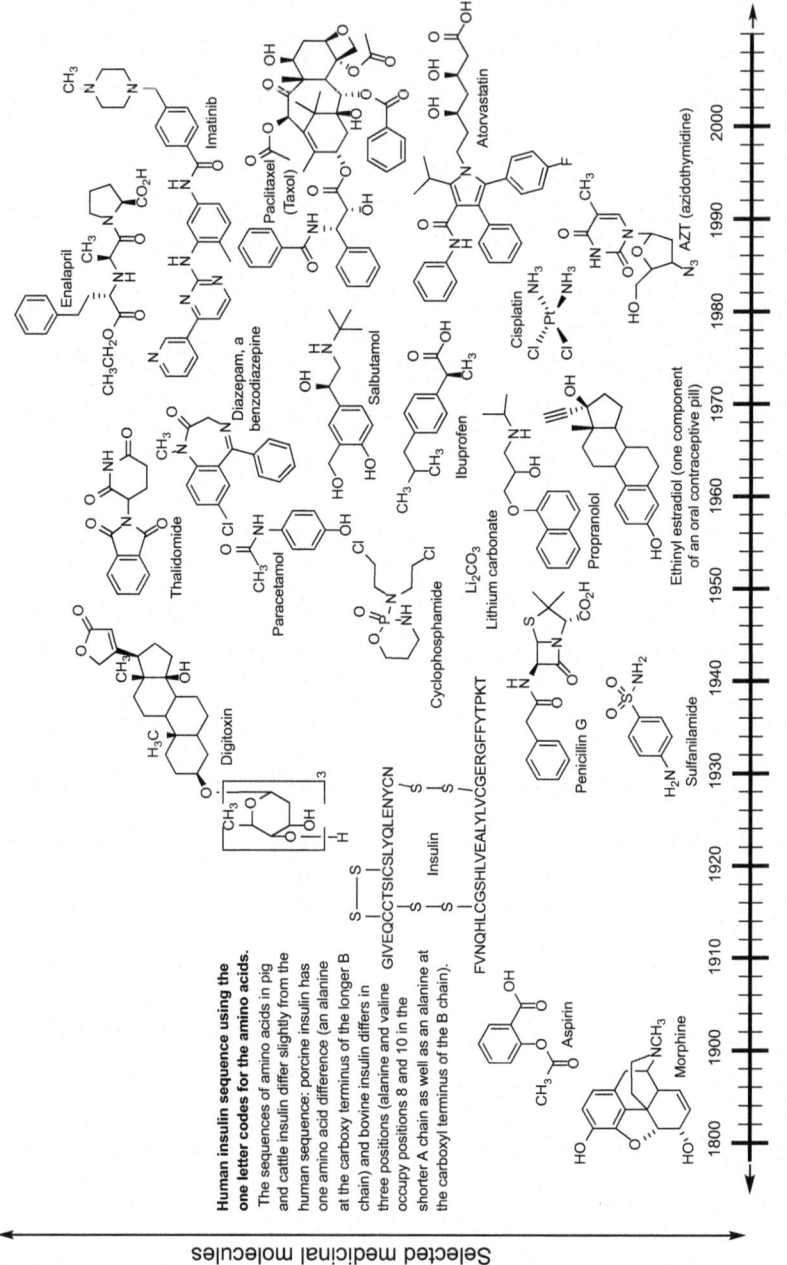

FIGURE 2.1 (*Continued*) (a) A selection of familiar drugs with an indication of the type of compound, condition or disease targeted and the date when they were introduced into clinical use. (b) Structures for the small molecule medicinal compounds given in Figure 2.1a.

However, once a promising medicinal molecule has been identified, it must success-fully pass through clinical trials and other regulatory steps before it can become a licensed therapeutic drug.

2.2 ASPIRIN AS A THERAPEUTIC COMPOUND

Aspirin (Figure 2.2) is one of the most commonly used medicinal molecules in the world today. The development of aspirin was one of the first examples in which syn-thetic chemistry was used to modify a lead compound. A lead compound is a substance that shows some interesting biological activity that is likely to be therapeutically use-ful. The lead compound which led to aspirin was salicylic acid (see Figure 2.3).

Preparations of the bark of the white willow tree (*Salix alba*) had been known for centuries to be effective in reducing pain and fevers, and it was the knowledge from this natural remedy that was the start of the series of events which culminated in the synthesis of aspirin (acetylsalicylic acid). An early written account of the antipyretic effects of willow bark was a letter to the Royal Society in 1763 from the Reverend Edward Stone in Oxfordshire, describing the use of willow bark infusions to success-fully treat agues (chills and fevers) in 50 persons.

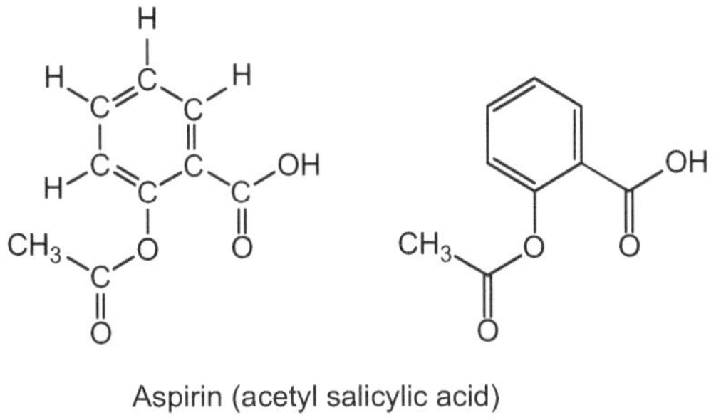

Aspirin (acetyl salicylic acid)

FIGURE 2.2 Alternative representations of aspirin structure. The structure on the left shows all the carbon, hydrogen and oxygen atoms whereas the structure on the right omits most of the carbon and hydrogen atoms. Hydrogen atoms are usually only shown in the skel-etal representation of organic structures when attached to a non-carbon atom. Selected hydro-gen atoms that are attached to carbon atoms may sometimes be shown to aid clarity, e.g. the CH$_3$ group in the skeletal structure of aspirin.

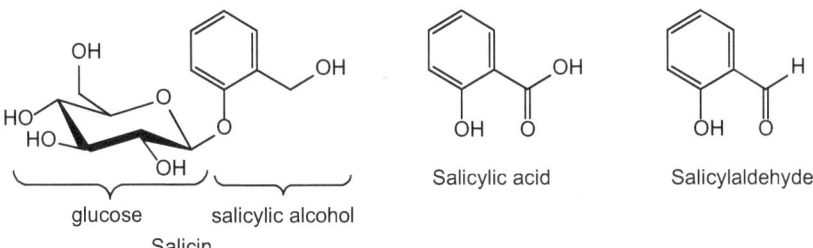

glucose salicylic alcohol

Salicin

Salicylic acid

Salicylaldehyde

FIGURE 2.3 Structures of key compounds in the discovery of aspirin.

The pharmacologically active ingredient within the willow bark extracts was first isolated, in an impure form, as a yellow powder in 1828 by Johann Andreas Buchner in Munich. This compound was called salicin, and the pure crystalline compound was subsequently obtained by the French pharmacist Henri Leroux in 1829 (Figure 2.3). The chemical structure of salicin ($C_{13}H_{18}O_7$) was established in 1859 by Hermann Kolbe at Marburg as a phenolic glycoside through a hydrolysis reaction to give D-glucose and salicylic alcohol. In 1876 the clinical effectiveness of salicin to treat rheumatic disorders of fever and joint pain was reported in the medical literature by the Scottish physician Thomas MacLagan. Salicin is, however, metabolised by the body to give salicylic acid and it is this compound that was later found to give the beneficial effects associated with willow bark extract. Parallel investigations in the 19th century with the leaves and flowers of the meadowsweet plant (*Filipendula ulmaria*) led to the isolation, by the Italian chemist Raffaele Piria working in Paris, of salicylaldehyde in 1839 which he then oxidised synthetically to produce salicylic acid. Initially, the therapeutic potential of salicylic acid was disappointing as it had an unpleasant taste and caused irritation to the stomach.

In the 1890s, the German chemical company Bayer investigated the medicinal properties of salicylic acid and sought to make an analogue that reduced these side effects. Salicylic acid is composed of a benzene ring with hydroxyl (OH) and carboxylic acid (COOH) groups; Felix Hoffmann of the Bayer company focused on modifying the phenolic OH. The acetylated derivative, acetylsalicylic acid, which Hoffmann synthesised in 1899 was found to be effective, had fewer undesirable side effects and is the compound now known as aspirin. Acetylsalicylic acid had first been prepared earlier in the century by Charles Frederic Gerhardt (1853) and Johann Kraut (1869), but these modified versions of salicylic acid were not used therapeutically or marketed. The name aspirin is derived from the old botanical name for meadowsweet, *Spiraea ulmaria*, which was the original synthetic source of salicylic acid and the prefix 'a' was added to indicate acetyl, the $CH_3-C=O$ group.

For many years aspirin was considered a prodrug for salicylic acid. Prodrugs are biologically inactive compounds which (after intake) are metabolised in the body to yield the active substance that is responsible for the drug's action. Aspirin was originally thought to be a prodrug for salicylic acid since it is rapidly transformed to this compound in vivo but work in the 20th century found that aspirin has its own innate biological activity.

The molecular basis for aspirin's analgesic and anti-inflammatory properties was identified in the 1970s. Aspirin was found to block the enzyme cyclooxygenase through an irreversible acetylation reaction (Figure 2.4). Cyclooxygenase catalyses the conversion of arachidonic acid to one of the prostaglandins, PGG_2, which in turn acts as a source for the synthesis of other prostaglandins (Figure 2.5). These

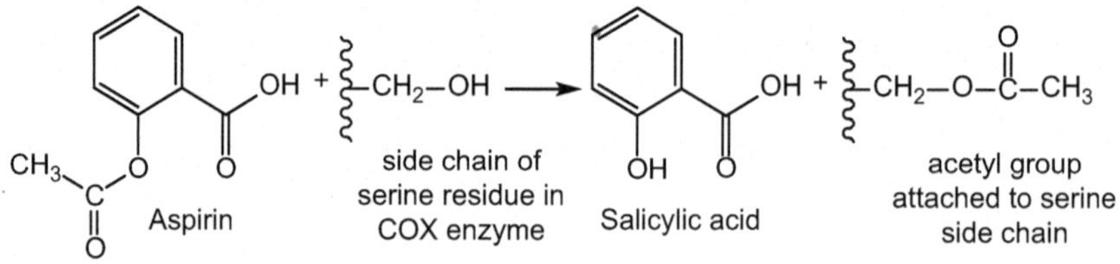

FIGURE 2.4 Aspirin as an enzyme inhibitor by acetylating a serine hydroxyl group at the active site of cyclooxygenase (COX).

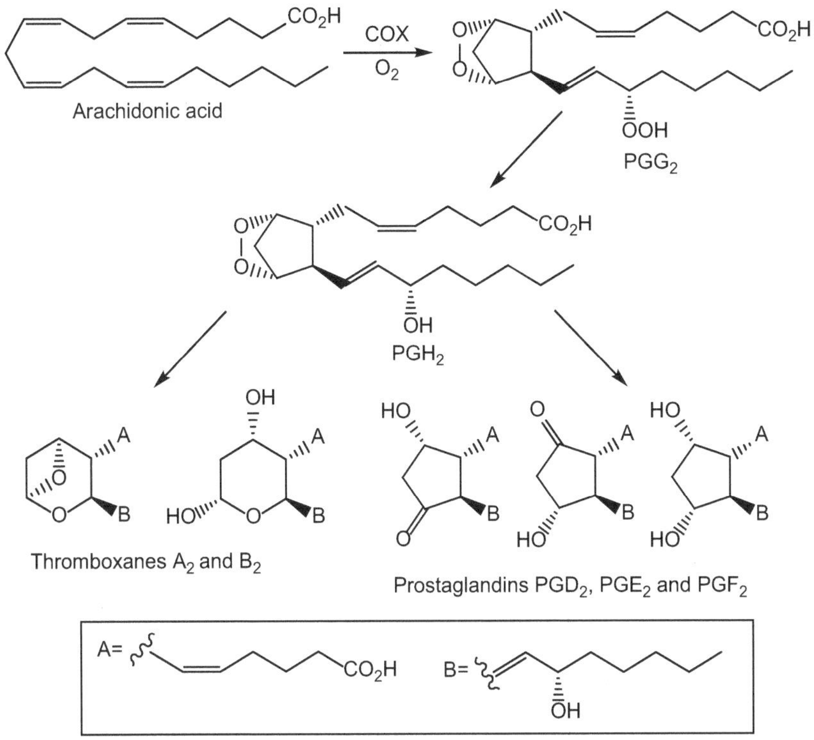

FIGURE 2.5 The structures of prostaglandins and related compounds produced in the body as a response to injury or disease.

lipid-like compounds are produced by most of the cells in the body when needed and are part of the body's way of dealing with injury and illness. Prostaglandins are made in excess at sites of tissue damage or infection where they cause inflammation, pain and fever as part of the healing process. By inhibiting the prostaglandin pathway, pain and inflammation are therefore suppressed. Furthermore, by irreversibly acetylating the active site of the cyclooxygenase enzyme, aspirin also inhibits the formation of the chemicals which cause the aggregation of blood platelets which is what starts a blood clot. As a consequence aspirin is now often involved in the treatment and prevention of cardiovascular disease.

2.3 KEY CONCEPTS

2.3.1 THE DRUG-BIOMACROMOLECULE INTERACTION

Drugs act by interfering with biological processes. The initial event involves chemical recognition between the drug and a biomacromolecule, such as a protein or nucleic acid. The biomacromolecule recognises the three-dimensional shape and arrangement of certain functional groups of the drug through intermolecular forces of attraction (Figure 2.6).

When a drug and receptor are sufficiently close the drug can dock into the receptor site via electrostatic interactions. Ionic bonds are the strongest of these non-covalent bonds; they are effective at distances greater than the other types of bonding and

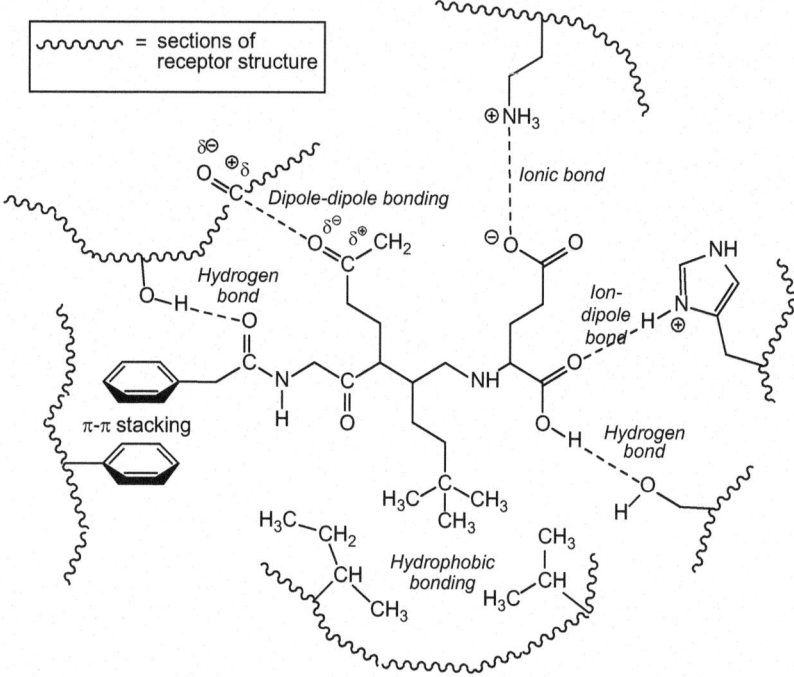

FIGURE 2.6   Hypothetical example of some of the common forms of intermolecular bonding found in drug-receptor interactions.

form first. The bonding is then strengthened by weaker electrostatic interactions in the form of ion-dipole, hydrogen bonds, dipole-dipole and aromatic-aromatic interactions. Hydrophobic bonding can occur when non-polar sections of molecules interact.

The drug-receptor interaction then either triggers a cascade of events finally resulting in the observed physiological change attributed to the drug, or blocks a biochemical step as exemplified above by the inhibition of cyclooxygenase by acetylsalicylic acid in the prostaglandin pathway. Drugs are most effective when their molecular geometry and electronic structure are complementary to those present at the receptor or enzyme site. The pattern of functional groups and substituents that are important for a drug's interaction with the biological target along with their relative positions in the molecule is known as the pharmacophore. Optimum requirements for receptor binding can demand precise positioning of a drug molecule's functional groups as was found during the design of ibuprofen.

Ibuprofen is, like aspirin, a non-steroidal anti-inflammatory drug and was developed in the mid-20th century as an alternative to aspirin. During the design and development of ibuprofen, a great many compounds were synthesised and tested for their effectiveness and side effects. Key structural elements were a carboxylic acid, an aromatic ring and a bulky alkyl group at position 4 of the ring. Structures of ibuprofen and three other promising candidates are shown in Figure 2.7. Each of these compounds contains the aforementioned structural triad, but it was the precise spatial relationship between the three groups in ibuprofen which led to this compound having an acceptable pharmacological profile, whereas the other compounds gave unacceptable side effects in clinical trials.

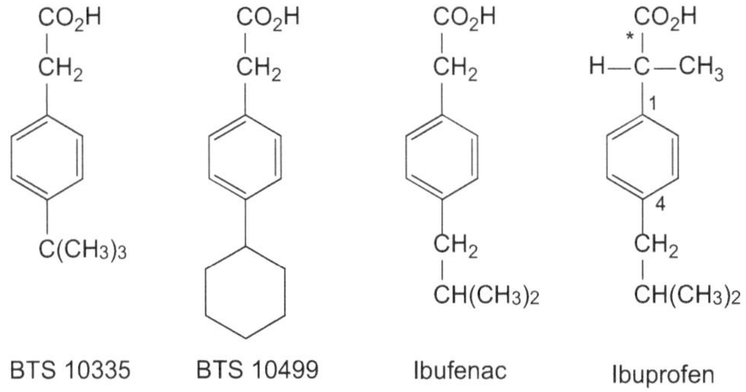

FIGURE 2.7   Some of the clinically tested compounds during the development of ibuprofen.

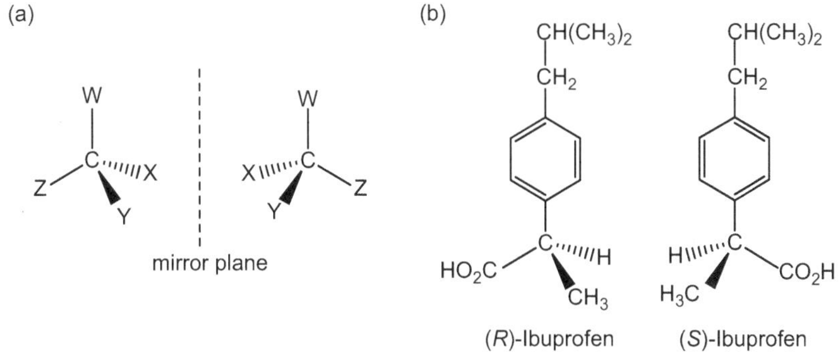

FIGURE 2.8   (a) Right- and left-handed forms shown as an object and its mirror image. (b) Right- and left-handed forms of ibuprofen.

Ibufenac was marketed in the UK but was later withdrawn due to liver toxicity in some patients. As with aspirin, ibuprofen is named from its component structural features: isobutyl (ibu), propanoic acid (pro) and phenyl (fen). Ibuprofen is a chiral compound and the commercial product is usually a racemic (1:1) mixture of two stereoisomers.

A chiral molecule is typically one which has at least one tetrahedral carbon atom with four different groups attached to it. The chiral centre in ibuprofen is marked with a star (*). Compounds with one chiral centre can exist as two non-superimposable mirror images which are called enantiomers (Figure 2.8). These two stereoisomers have different shapes and, if they interact with chiral biological macromolecules, they may exhibit different biological activities. The binding of an enantiomer to a chiral receptor is analogous to a hand fitting into a glove, with the glove acting as the receptor. Your left and right hands are mirror images and your left hand could fit a left-hand glove perfectly but not fit into the corresponding right-hand one. The 'right hand' and 'left hand' forms of chiral centres are designated as '*R*' and '*S*', from the Latin *rectus* and *sinister*, respectively, based on a set of rules known as the Cahn-Ingold-Prelog system. The difference in biological action between stereoisomers can be considerable, with one enantiomer showing a beneficial physiological effect whilst the other may be less active, inactive, give unwanted side effects or be harmful.

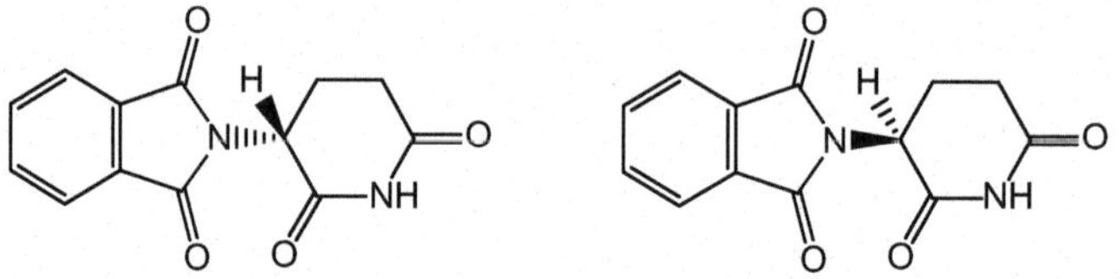

(R)-Thalidomide   (S)-Thalidomide

FIGURE 2.9   *R* and *S* forms of thalidomide; the *S*-isomer has teratogenic activity.

Most commercial formulations of ibuprofen are the racemic mixture as it is difficult to separate the two enantiomers. (*S*)-Ibuprofen has considerably greater anti-inflammatory activity than (*R*)-ibuprofen but the inactive *R*-form is converted into the biologically active form within the body.

Following the thalidomide tragedy of the late 1950s and early 1960s changes were made to the way drugs were tested, approved and marketed. Thalidomide was introduced in the late 1950s to treat morning sickness in pregnant women. However, it was found to cause severe birth defects in children whose mothers had taken the drug during their first trimester of pregnancy, and the drug was banned and withdrawn in 1961/2. The thalidomide molecule has a chiral centre and was sold in the late 1950s as the racemate. However, after the tragedy the two enantiomeric forms of the compound were found to have different biological activities; the *R*-enantiomer was effective against morning sickness whilst the *S*-enantiomer was teratogenic and induced abnormalities in human embryos (Figure 2.9). However, both the *R* and *S* enantiomers can rapidly interconvert (racemise) in bodily fluids and in water, forming equal concentrations of each form. Both enantiomeric forms of a prospective new chiral drug are now synthesised and the pharmacological profiles of each individual chiral species are tested. Medications can no longer be approved purely on the basis of animal testing.

Thalidomide has re-emerged in the 21st century as a treatment for leprosy and as an anticancer agent. The mechanism of antitumour action is complex but thalidomide is thought to interfere with the process that leads to the formation of new blood vessels from pre-existing ones; it has been suggested that the thalidomide-induced damage to a rapidly developing embryo may be linked to an anti-angiogenesis effect. Despite the tight regulations over the prescription of thalidomide, a small group of children have sadly been born with limb deformities in Brazil in recent years. The continuing tragedy appears to result from the use of thalidomide to treat complications of leprosy, such as skin lesions, in association with a local cultural practice of sharing medicines.

2.3.2   GENERAL PRINCIPLES OF DRUG DESIGN AND DEVELOPMENT

The short linear sequence from willow bark (folklore remedy) → salicin (active agent) → salicylic acid (lead compound) → aspirin (licensed drug) is an illustration of how an existing molecule informs the design of future molecules. At each step of this process, it was a combination of the scientists' knowledge, experience and intuition, and to some extent luck, that resulted in a synthetic compound which met a medical

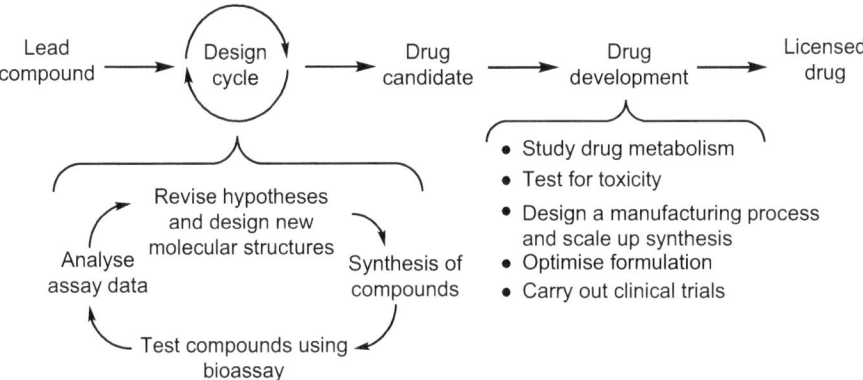

FIGURE 2.10   Key events in the drug discovery, design and development process.

need. A number of the drugs in Figure 2.1, including ibuprofen, were developed by a similar process but with a more involved and iterative design cycle (Figure 2.10). In every design cycle, the method either finds a molecule better than that already known or fails to do so. Either result extends the knowledge of the structural and electronic factors governing the activity for all the properties assayed. In the case of aspirin, acetylsalicylic acid was both the drug candidate and licensed drug and very little development phase occurred. In modern drug discovery programmes, this phase can be prolonged, particularly the clinical trials component. After the promising candidate molecules are patented, they are extensively tested under laboratory conditions to obtain efficacy and safety information before entering clinical trials in humans to establish the safety and efficacy in situ. The original method of making the drug candidates often requires modification or is completely altered to enable large-scale production. Many promising compounds fail at each stage of a drug discovery programme, thus requiring the synthesis and testing of further analogues.

A drug discovery programme may take 10 or more years with upwards of 10,000 compounds synthesised of which only a handful of molecules may survive to enter clinical trials. The entire process can be thought of entering and navigating a maze where there are many dead ends and wrong turns but with no guarantee of an exit, i.e. a successful product. It is a very expensive process, costing hundreds of millions of pounds and a pharmaceutical company has only a relatively short period of time, until the patent expires, to recoup the cost of their discovery and make a profit. A patent gives a company exclusive rights to manufacture and sell a new pharmaceutical for 20 years but this is from the time of application when new compounds have been made and found to have significant activity, not when the drug comes onto the market.

New technologies are helping to accelerate some of the lengthy steps in drug discovery and make it more efficient. For example, high-throughput screening of compound libraries can identify if any existing molecules have activity against the disease of interest, thereby giving new lead compounds. Parallel and combinatorial methods can speed up the synthesis of new compounds and computer modelling can help determine the fit and binding of a compound to the receptor site of a biomacromolecule. The crystal structure and modelling of a specific oncoprotein was

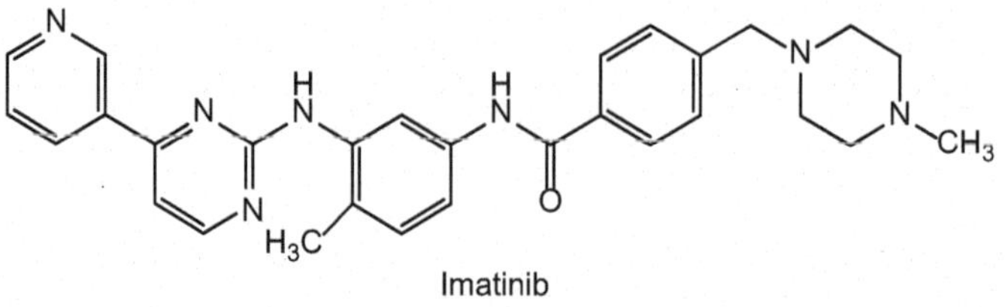

FIGURE 2.11 Structure of imatinib, a drug used to treat leukaemia, and sold under the brand name Gleevec.

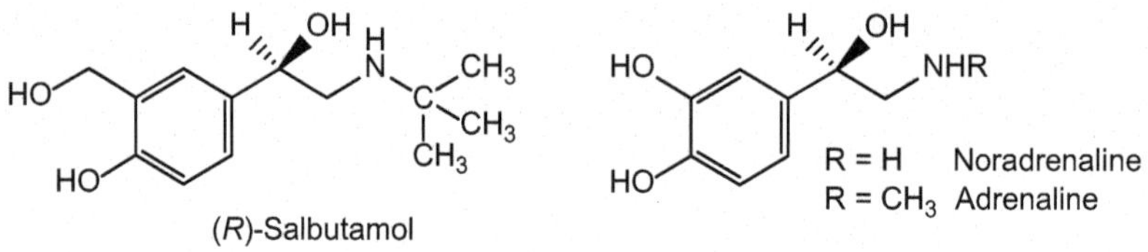

FIGURE 2.12 Structures of salbutamol and naturally occurring lead compounds.

pivotal in optimising the design and development of imatinib, a drug treatment for chronic myeloid leukaemia (Figure 2.11). Proteomics and genomics are also now giving greater insight and understanding of molecular responses to cell and tissue damage such as how cell-surface membrane proteins mediate host-pathogen interactions.

The discovery of aspirin was largely carried out by individuals, whereas modern drug discovery is a team effort involving a range of specialists such as chemists, biologists, pharmacologists, toxicologists, molecular modellers, clinicians, statisticians and patent lawyers. Despite technological advances, the close relationship between chemical structure and physiological action remains an unchanging element in the search for new medicinal molecules.

2.3.3 ASPIRIN AS A STARTING MATERIAL: SYNTHESIS OF SALBUTAMOL

Salbutamol is used in the treatment of asthma where it acts as a bronchodilator. Its therapeutic effect is due to the selective stimulation of β_2-adrenergic receptors which relax the bronchial smooth muscle, thus allowing dilation, or widening, of the airways. Salbutamol was developed from the modification of noradrenaline, a natural chemical messenger with a myriad of roles in the body (Figure 2.12). Adrenaline, the structurally related compound with an additional methyl group on the nitrogen atom, was the lead compound for the class of molecules collectively known as β-blockers or β_1-adrenergic receptor antagonists, which have an important role in the treatment of cardiovascular disease (see Section 2.5).

When planning how to make a target molecule such as salbutamol, the synthetic chemist has a number of factors to consider. This list, albeit not comprehensive, includes the reaction conditions, rate and yield for each step, safety such as the toxicity and flammability of reagents, purification methods and the financial economy of the overall process. Salbutamol can be prepared from a number of cheap and readily available reaction materials but the scheme shown in Figure 2.13 reflects the original

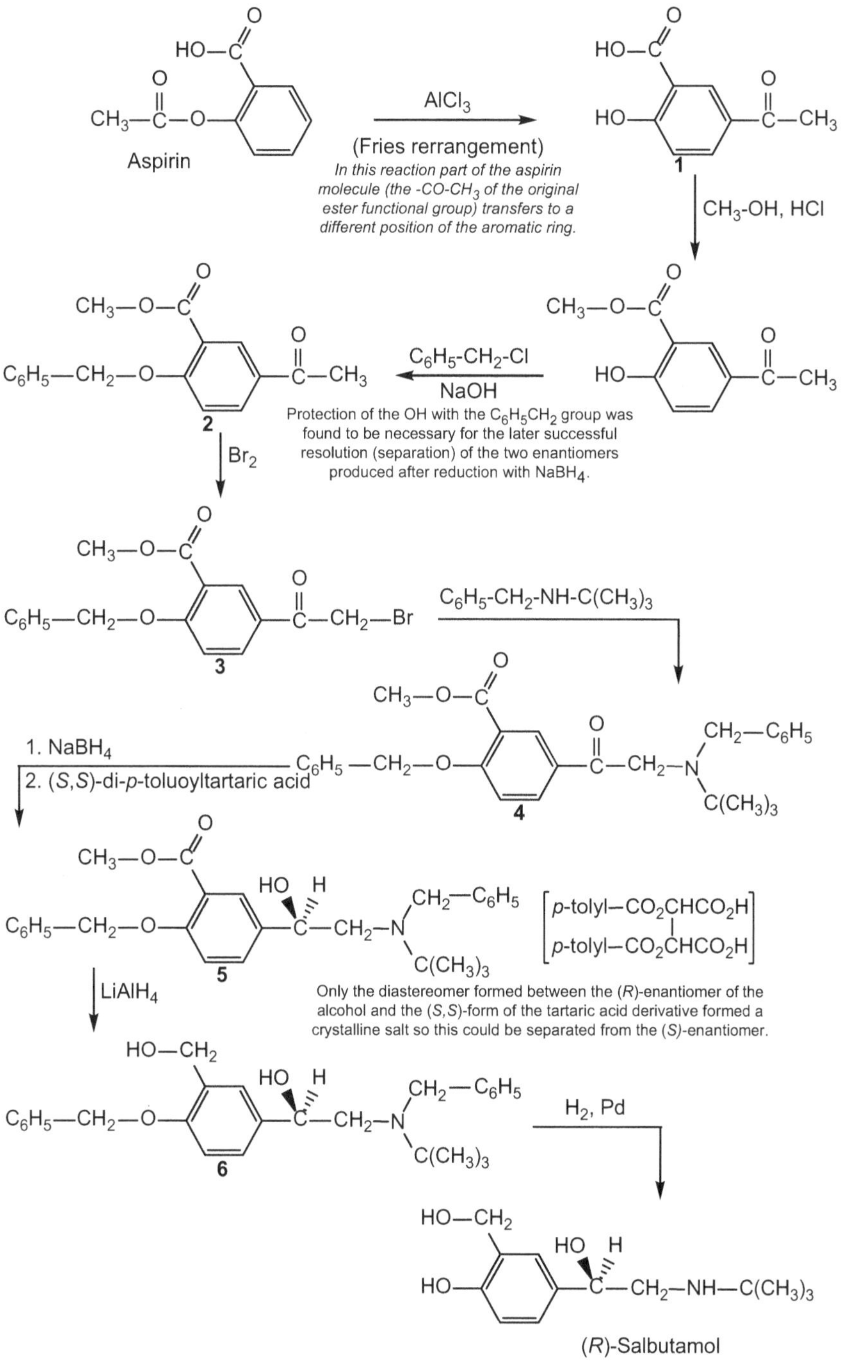

FIGURE 2.13 Synthesis of (R)-salbutamol from aspirin. Intermediates in the synthesis referred to in the text are numbered **1** to **6**. Note that p-tolyl is a functional group related to toluene, formed by the removal of a hydrogen atom from the ring carbon at the 4-position and has the structure –C_6H_4–CH_3.

laboratory synthesis. The starting material is acetylsalicylic acid and the synthesis involves the separation of a racemic mixture to give the two enantiomers.

The first step in the synthesis is a Fries rearrangement of aspirin in which the acetyl group, –$COCH_3$, of the phenolic ester migrates to the opposite position of the aromatic ring and the product **1** has hydroxyl and ketone groups in addition to the original carboxylic acid. The carboxylic acid was then esterified and the phenolic OH group protected as the benzyl ether, giving compound **2**. This latter step was crucial to the successful resolution procedure further in the synthetic route. Reaction of the aromatic ketone **2** with bromine gave the bromoketone **3** which was then converted to the aminoketone **4** with *N*-(*tert*-butyl)benzylamine. Reduction of both the ketone and ester carbonyl groups in compound **4** could be accomplished in one step with lithium aluminium hydride, $LiAlH_4$, or in two separate steps. This latter method was advantageous as the racemic alcohol **5** produced from the reduction with sodium borohydride, $NaBH_4$, could be resolved into the (separated) component enantiomers through a preferential crystallisation procedure.

The racemic mixture **5** obtained from the reduction step with $NaBH_4$ was reacted with an enantiomerically pure chiral reagent, in this case a derivative of tartaric acid. In this acid-base reaction, a proton (H^+) is transferred from the chiral acid reagent to the tertiary amine to give a mixture of diastereomeric salts which could be separated (Figure 2.14). Only the diastereomer formed between (*S,S*)-di-*p*-toluoyltartaric acid and compound **5** with *R*-stereochemistry was insoluble in ethyl acetate and was purified through recrystallisation and subsequently converted into the pure enantiomer of **5**. The *S*-enantiomer was obtained in a similar procedure using (*R,R*)-di-*p*-toluoyltartaric acid. The ester functionality of **5** was then reduced with $LiAlH_4$ to give compound **6**, and the two benzyl groups were removed using catalytic hydrogenolysis to afford the enantiomerically pure (*R*)-salbutamol.

The *R*-isomer was 68 times as active as the *S*-isomer as a β_2-stimulant and was developed into the familiar bronchodilator drug. Moreover, salbutamol has very low potency on cardiac β_1-adrenergic receptors which was another key requirement of the original brief: develop a compound that works as a bronchodilator whilst minimising detrimental cardiovascular interactions. Other effective salbutamol analogues have subsequently been prepared. Salmeterol, for example, with a long chain 4-phenylbutoxyhexyl substituent instead of the *tert*-butyl group on the nitrogen atom, is

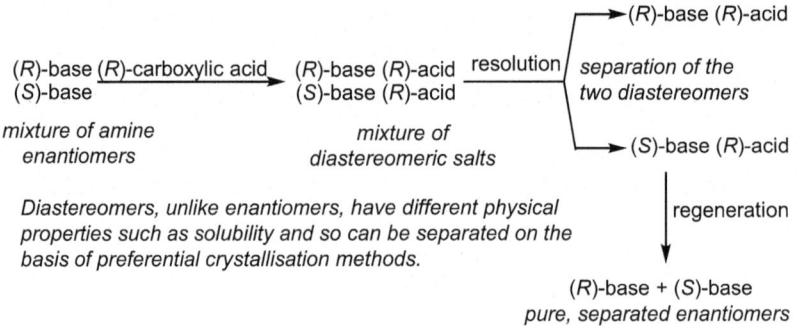

FIGURE 2.14   Enantiomeric resolution via diastereomers using a chiral acid as the resolving agent.

more than twice as potent as salbutamol and has a longer duration of action, 12 hours compared to 4 to 6 hours for salbutamol.

2.4   OPIOID ANALGESICS

2.4.1   MORPHINE, CODEINE AND DIAMORPHINE

Preparations of the resin extracted from the seed capsule of the opium poppy, *Papaver somniferum*, have been used therapeutically for thousands of years, initially largely as a sedative and treatment for coughs and later as a means to alleviate pain. Laudanum, a 10% solution of opium powder in alcohol, for example, was widely used in the Victorian era as a remedy for everything from pain and insomnia to 'female disorders' and melancholy.

The term 'opioid analgesics' or 'opiates' refers to the family of compounds, naturally occurring, synthetic or semi-synthetic based on the structure of morphine, the principal opium alkaloid. Morphine was first isolated and purified by the German pharmacist Friedrich Sertürner in the early 19th century and became commercially available in the 1830s. The name morphine is derived from Morpheus, the Greek god of dreams, and the 'ine' ending indicates that the compound is an amine. The sleep-inducing properties of opiates led to these compounds also being referred to as narcotics, from the Greek word *narkotikos*, 'to make numb', although this term often now refers to the illegal use of drugs.

The functional groups in morphine had been identified by the 1880s but the complete structure of the molecule (see Table 2.1) was not finally elucidated until 1925. The Bayer company started production of diacetylmorphine (heroin) at the end of the 19th century, co-advertising this compound as 'the sedative for coughs' alongside that for aspirin as a 'substitute for the salicylates, agreeable of taste, free from unpleasant after effects'. Although aspirin and opiates are both used in the relief of pain, they have different mechanisms of action; aspirin interferes with the synthesis of prostaglandins, whilst opiates work in the central nervous system (CNS) by binding to pain-reducing receptors in the brain. Receptor binding also leads to the undesired side effects of the opiates such as respiratory depression and addiction. Heroin addiction rates rose alarmingly in the early 20th century and the Dangerous Drugs Act of 1920 led to stricter controls over the use of opium and opiates in the UK.

Due to the complex nature of the morphine structure (five rings, five chiral centres and five functional groups), legally grown and harvested opium poppies are the principal source of this compound, and also of thebaine, one of morphine's biological precursors, and the 3-*O*-methyl ether derivative, codeine. Whilst codeine is present in small amounts in opium, most of the pharmaceutical material is manufactured by methylation of morphine. Thebaine itself has very little intrinsic analgesic activity and its main value is as a starting material for a number of semi-synthetic drugs.

The differences in biological activity between morphine, codeine and diamorphine lie in the pharmacokinetics of these compounds, i.e. the **A**bsorption, **D**istribution, **M**etabolism and **E**limination processes rather than the intrinsic affinity for the receptor. The ADME processes influence how a drug moves through the body to reach its site of action in sufficient concentration to enable the pharmacological response to occur. The opiates are given orally or by injection and to exert their biological

TABLE 2.1
Opiates and Their Analgesic Activity Compared to Morphine and Some of the More Common Medicinal Applications

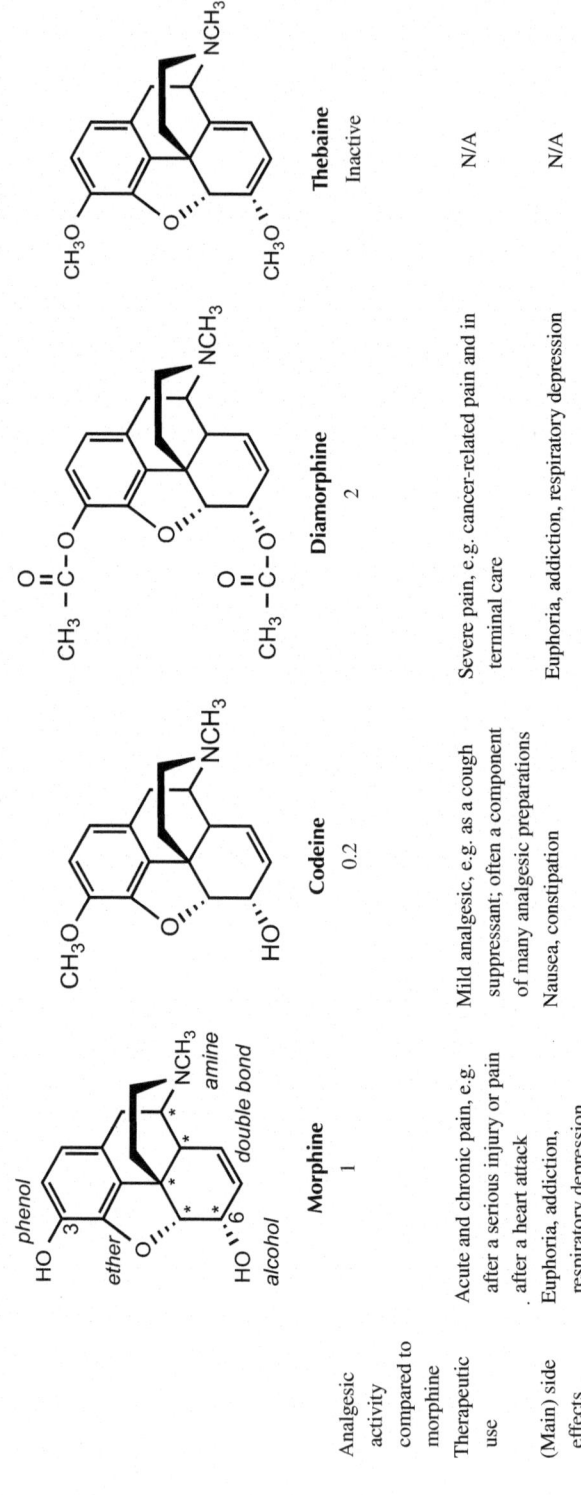

	Morphine	Codeine	Diamorphine	Thebaine
Analgesic activity compared to morphine	1	0.2	2	Inactive
Therapeutic use	Acute and chronic pain, e.g. after a serious injury or pain after a heart attack	Mild analgesic, e.g. as a cough suppressant; often a component of many analgesic preparations	Severe pain, e.g. cancer-related pain and in terminal care	N/A
(Main) side effects	Euphoria, addiction, respiratory depression	Nausea, constipation	Euphoria, addiction, respiratory depression	N/A

The structure of morphine indicates the functional groups, the position of chiral centres (marked by an asterisk) and relevant numbering. Diamorphine is synonymous with the names diacetylmorphine and heroin and the structures of these compounds are identical. The name heroin is more commonly associated with its use as an illegal drug substance where it is unlikely to be a pure compound. A serious side effect of morphine and diamorphine is breathing difficulties due to the activation of opioid receptors in respiratory control centres in the brainstem. Respiratory depression can be lethal and suffocation is the most common cause of death from overdose or abuse of morphine and heroin. An opioid antagonist such as naloxone (Figure 2.18) has a role in the treatment of overdose in abuse and clinical settings. An antagonist binds to the receptor effectively displacing the morphine or other opioid agonist from binding and reverses the respiratory depression and analgesic effects of the opioid drug. Addiction and physical and psychological dependence are also adverse side effects of opioid use with serious consequences.

activity they must travel to the receptors in the brain. In order to do this the molecules must pass through the blood-brain barrier, a lipophilic membrane. This barrier acts as a self-defence mechanism to prevent the passage of potentially harmful substances from the blood into brain tissue. Increasing the lipophilicity of a compound can improve blood-brain barrier permeability and give a more effective pharmaceutical.

Codeine has a methyl group on the OH group at position 3, making this a more lipophilic (fat-like) compound compared to morphine with two polar OH groups. However, as codeine is administered orally it enters the bloodstream through the gastrointestinal tract and then passes through the liver where it is partially demethylated and transformed into morphine. Codeine can therefore be considered a prodrug for morphine and its analgesic properties are largely dependent upon its conversion in the body to morphine. Codeine has 20% analgesic activity of morphine as only a small percentage (5%–10%) of it is metabolised to the active molecule with the vast majority of an administered dose being converted into inactive metabolites and excreted.

Morphine has three polar groups: the phenolic OH, the OH at position 6 and the amine. Both of the OH groups are masked by acetyl groups in diamorphine which results in a more lipophilic compound. Consequently, diamorphine is more effectively absorbed into the brain but it also has a lower number of hydrogen bond donors in the structure which makes it better able to evade the efflux mechanisms of the blood-brain barrier that shuttle molecules back out of the brain. Diamorphine is usually administered by injection and is deacetylated by esterases in the brain to 6-acetylmorphine and morphine, both of which bind to the opioid receptors. Diamorphine has twice the analgesic activity of morphine, a faster onset of action and can also be considered a prodrug of morphine.

2.4.2   THE OPIOID PHARMACOPHORE

Morphine has five functional groups: a phenolic OH, an alcohol, an ether bridge, a double bond and a tertiary amine. Which of these are required for biological activity and which are not was established using structure-activity relationship (SAR) studies. Compounds were prepared with minor changes to the lead compound (morphine) and the effect that this had on the analgesic activity was assessed. The functional groups in morphine, and their relative positions, which were found to be essential for binding to the receptor and thus elicit analgesic activity were the aromatic ring, the phenolic OH group at position 3 and the nitrogen atom, protonated at physiological pH (Table 2.2). The biological importance of the phenolic OH group had already been established from the studies with codeine, diamorphine and other similar analogues (Table 2.1).

Morphine has a rigid T-shaped structure and the pharmacophore can be defined as a triangle where each corner of the triangle represents one of the functional groups (Figure 2.15).

2.4.3   SIMPLIFICATION OF THE MORPHINE STRUCTURE

Morphine is an excellent, but not ideal, painkiller due to the serious side effects such as respiratory depression and addiction. Many hundreds, if not thousands, of morphine analogues have been synthesised over several decades in the search for an

TABLE 2.2

Relative Analgesic Potency of Some Morphine Analogues

Compound	R_1	R_2	Analgesic Activity Compared to Morphine
Morphine	H	H	1
Heterocodeine	H	CH_3	5
3-Acetylmorphine	$CH_3-\overset{\overset{O}{\|\|}}{C}$	H	< 0.1
6-Acetylmorphine	H	$CH_3-\overset{\overset{O}{\|\|}}{C}$	4

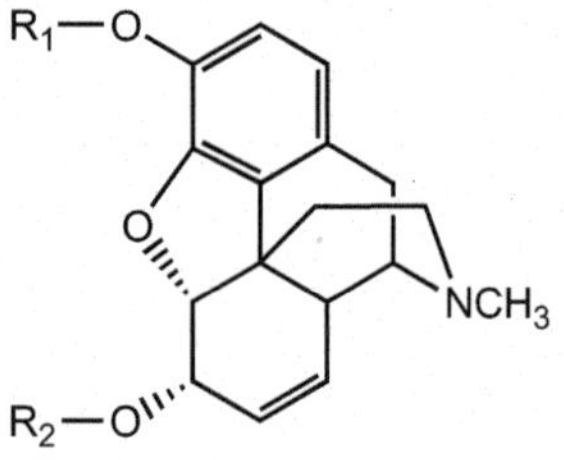

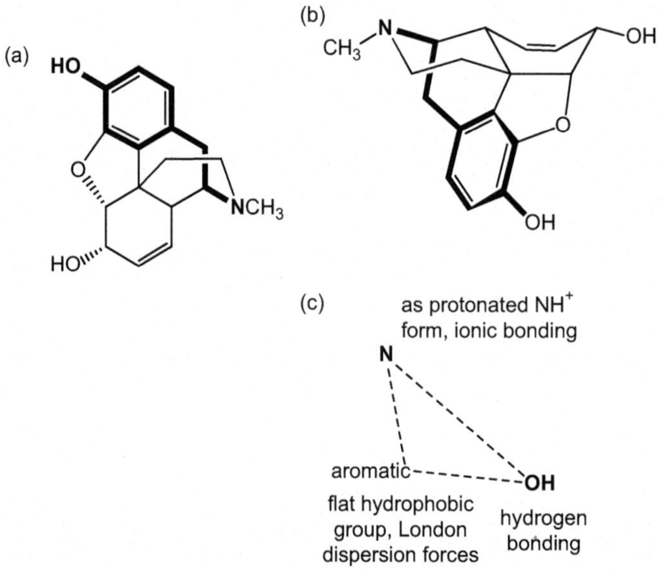

FIGURE 2.15 (a) Morphine with pharmacophore shown in bold type. (b) T-shape structure of morphine with the pharmacophore in bold. Two of the rings are roughly at right angles to the other three and form the crossbar of the **T** with the phenol OH and aromatic ring protruding forming the stem of the **T**. The N–CH–CH component is part of two rings and holds the molecule in a particular orientation. (c) Possible binding interactions between pharmacophore of morphine and receptor.

effective analgesic devoid of dangerous side effects. Only a fraction of the morphine structure is the pharmacophore, so one strategy centred around removing non-essential parts of the molecule which would result not only in a simpler compound with fewer chiral centres, but one which would be easier and cheaper to synthesise.

Moreover, reducing the molecular complexity was expected to give compounds with more favourable physicochemical profiles.

Systematically reducing the complexity of the ring system led to a number of marketed drugs, as illustrated by the four synthetic compounds in Figure 2.16, along with some salient information on their biological activity. The five rings in morphine are labelled **A** to **E** and omission of the furan ring, ring **D**, gives a morphinan structure. Levorphanol is five times as potent as morphine, demonstrating that not only was the oxygen bridge not a critical binding element but was probably impeding interaction with the receptor. Moreover, this compound also further illustrated the non-essential nature of the polar OH group in ring **C**. Removing both rings **C** and **D** keeps the pharmacophore intact and leads to the 6,7-benzomorphan series of compounds. Pentazocine represents this class of opioids and two methyl groups were retained in ring **B** to indicate ring **C** and a dimethylallyl group was substituted for the methyl group on the nitrogen. Replacing the methyl group on the nitrogen with other alkyl substituents was a way of probing for additional potential binding sites at the receptor with a view to making opiates free from respiratory depression but unexpectedly led to the discovery of antagonist effects. Pure opioid antagonists such as naloxone (Figure 2.18) are used in the treatment of opioid overdose and the reversal of respiratory depression associated with opioid use (see Table 2.1 legend).

Pentazocine is a mixed agonist-antagonist through binding to three of the four opioid receptors; it acts as an agonist at the delta (δ) and kappa (κ) receptors and an antagonist at the mu (μ) receptor. Later work established that key clinical effects of opiate drugs are mediated by subtle differences in binding to the receptor subtypes. Pentazocine is an orally active analgesic used to treat moderate to severe pain but high doses elicit psychotomimetic effects such as hallucination and dysphoria which limit its clinical applications.

The 4-phenylpiperidine derivatives represent the ultimate simplification of the morphine skeleton whilst retaining the key pharmacophore components of the aromatic ring (**A**) and the piperidine ring (**E**) containing the amine group but not the 3-OH group. Pethidine was discovered by chance in 1939 during a search for synthetic substitutes for atropine and was found to possess analgesic properties in addition to antispasmodic effects. Pethidine is a widely used synthetic opioid in clinical practice, particularly during childbirth, despite its addictive properties. Although opioids can cross the placenta the short duration of action of pethidine means that there is less chance of the drug causing problems with the baby's breathing. From a structural viewpoint pethidine is a model drug molecule being straightforward to synthesise (two steps from benzyl cyanide), inexpensive and has no chiral centres. The compound, 3-hydroxypethidine, with all three components of the pharmacophore, is slightly more potent as an analgesic than pethidine but has all the drawbacks of other opioids and is not of any clinical significance. Therefore, it appears that an aromatic ring and nitrogen atom in the correct spatial relationship is an essential feature for the analgesic pharmacophore but the phenol OH group is not.

Fentanyl (see Figure 2.18) was designed along similar lines to the 4-phenylpiperidines and has 100 times the potency of morphine due to its high lipophilicity and excellent transport across the blood-brain barrier. However, it is highly addictive and

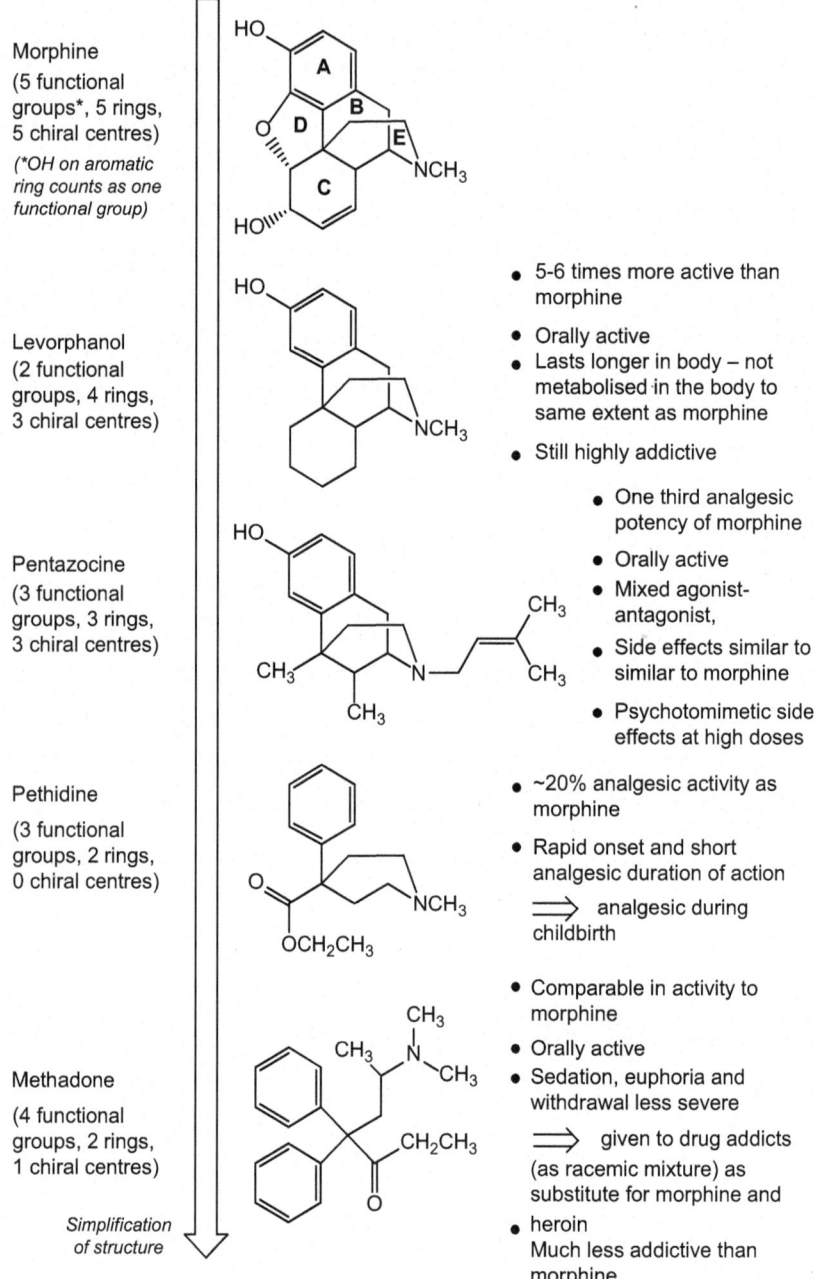

Morphine
(5 functional
groups*, 5 rings,
5 chiral centres)

(*OH on aromatic
ring counts as one
functional group)

Levorphanol
(2 functional
groups, 4 rings,
3 chiral centres)

- 5-6 times more active than
  morphine
- Orally active
- Lasts longer in body – not
  metabolised in the body to
  same extent as morphine
- Still highly addictive

Pentazocine
(3 functional
groups, 3 rings,
3 chiral centres)

- One third analgesic
  potency of morphine
- Orally active
- Mixed agonist-
  antagonist,
- Side effects similar to
  similar to morphine
- Psychotomimetic side
  effects at high doses

Pethidine
(3 functional
groups, 2 rings,
0 chiral centres)

- ~20% analgesic activity as
  morphine
- Rapid onset and short
  analgesic duration of action
  $\Longrightarrow$ analgesic during
  childbirth

Methadone
(4 functional
groups, 2 rings,
1 chiral centres)

- Comparable in activity to
  morphine
- Orally active
- Sedation, euphoria and
  withdrawal less severe
  $\Longrightarrow$ given to drug addicts
  (as racemic mixture) as
  substitute for morphine and
- heroin
  Much less addictive than
  morphine

Simplification
of structure

FIGURE 2.16   Structural simplification of morphine leading to marketed drugs.

the use of fentanyl has contributed to the dramatic rise in drug overdoses reported in
the USA particularly when obtained from illegal sources.

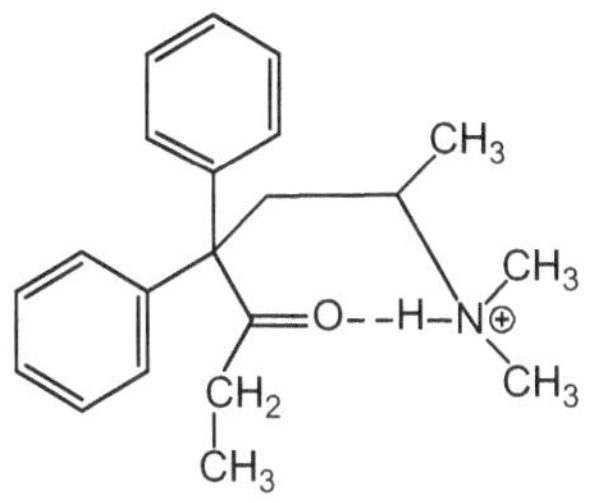

FIGURE 2.17 Methadone drawn to show the virtual seven-membered heterocyclic ring through the formation of an intramolecular hydrogen bond involving the protonated tertiary amine.

The piperidine ring system **E** is no longer present in methadone. This diphenyl-propylamine molecule appears at first to have very little structural similarity to morphine. However, a methadone conformation involving a 'virtual heterocyclic ring' which occupies the same space as the piperidine ring of morphine can be envisaged through the formation of an intramolecular hydrogen bond (Figure 2.17). The methadone molecule has been extensively modified without yielding any major advantage over this parent compound. Methadone is occasionally used for pain relief but is more widely used as a substitute drug for people addicted to opiates, particularly heroin.

2.4.4 Introduction of New Groups to the Morphine Structure

Another strategy adopted in the process of designing and developing new pharmaceuticals is the introduction of additional groups into the lead compound. The basis for this strategy is that more complex molecules may provide more potential groups for interaction with a receptor binding pocket and/or alter physicochemical properties such as lipophilicity of the molecule which may impact on membrane transport. Longer alkyl groups, e.g. $CH_3–CH_2–CH_2–$, increase the lipophilicity of a molecule whereas additional OH groups increase the polarity of a compound and lower the lipophilicity. If additional groups make a compound more rigid then it may restrict the flexibility of the molecule and make it more likely to be in the active conformation for binding to the target receptor and/or reduce the likelihood of interaction with unwanted receptors.

Substitution of the CH_3 group on the nitrogen atom on morphine with the allyl group, $CH_2–CH=CH_2$, increased the lipophilicity of the molecule but the compound (nalorphine) had less of an analgesic effect than morphine. However, it was found to have antagonist effects and was the first opioid marketed as an opioid reversal agent. Opioid antagonists occupy the same receptor sites as the agonists but block the pharmacological effect. Moreover antagonists can displace opioid drugs from the receptor sites, thus reversing their effects. Nalorphine was superseded by naloxone (Figure 2.18) which has greater receptor affinity and is a pure antagonist. Naloxone is used to treat opiate poisoning, including in children born to heroin addicts.

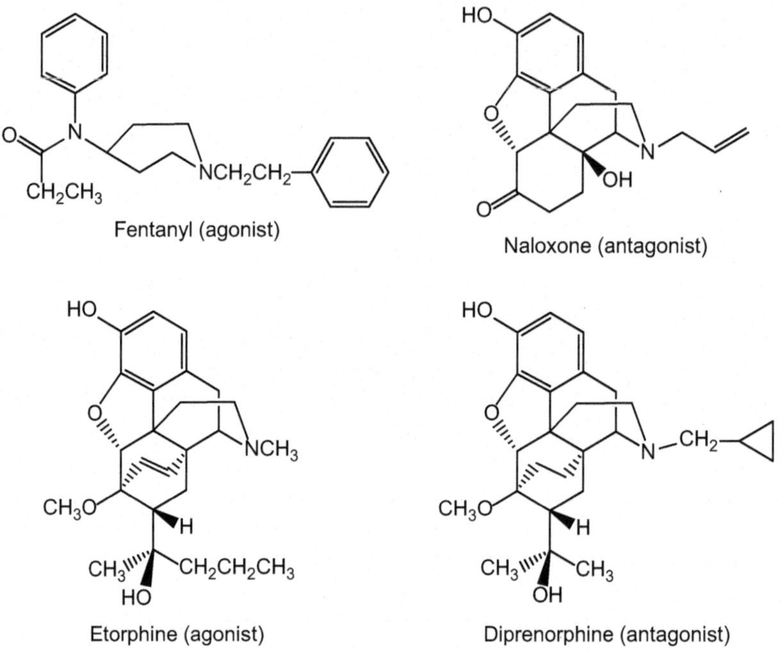

Fentanyl (agonist)

Naloxone (antagonist)

Etorphine (agonist)

Diprenorphine (antagonist)

FIGURE 2.18   Structures of various opioid agonists and antagonists.

Etorphine has an extra ring system built onto the cyclohexanol (ring **C**) which increases the rigidity of the opiate structure. The bridged opiate structure is formed through an initial Diels-Alder reaction between the conjugated diene system of thebaine with an unsaturated ketone followed by a Grignard reaction on the ketone and then hydrolysis of the 3-methoxy group to give the free phenolic OH. Etorphine has 5,000–10,000 times the analgesic potency of morphine due to a combination of its great lipophilicity and ease of penetration of the blood-brain barrier (300 times) together with better binding and higher affinity for the receptor (20 times). This highly potent analgesic is most widely used in veterinary medicine to immobilise large animals such as elephants and rhinoceros in the wild and in zoos by means of tranquiliser darts. Veterinary strength etorphine is fatal to humans and so an opioid antagonist such as naloxone must be readily at hand in case of accidental self-administration. Diprenorphine is a specific antagonist to etorphine and is used to remobilise animals and aid their recovery from the anaesthetic procedure.

2.4.5   WHY MORPHINE WORKS AS AN ANALGESIC: THE ENDOGENOUS LIGANDS

Morphine is part of the defence mechanism of the opium poppy and so would not naturally be expected to bind to receptors on nerve cells in the human brain. The fact that it does is a fortunate happenstance because the structure of morphine has a resemblance to that of endogenous molecules which are naturally produced in the human body. By being able to occupy the same receptor site in the brain as the endogenous ligands, the opiate drugs thus mimic their effects in suppressing the transmission of pain impulses within the nervous system.

The receptors for the endogenous compounds were discovered in the 1970s from studies involving mice and radiolabelled levorphanol which identified their location to within the brain. Subsequent work has shown opioid receptors to be widely present within the CNS and to a lesser extent throughout the peripheral tissues. The 'endogenous morphines' were discovered shortly thereafter in brain extracts and were identified as two pentapeptides named enkephalins and the longer peptides β-endorphin and the dynorphins which contain the enkephalin sequences at their N-termini (Table 2.3). These opioid peptides are now known to be cleaved from larger polypeptide precursors. Evidence for the existence of multiple subtypes of receptor came from studies involving several analgesic agents and were named μ or mu (after morphine), κ or kappa (from ketocyclazocine) and δ or delta (after vas deferens, the tissue within which it was first isolated).

All of the **end**ogenous **m**orphine peptides, often collectively known as the endorphins, share one of the enkephalin sequences at the N-terminus, and are produced during physical exercise, thus elevating the pain threshold and they may also contribute to the positive mood effects associated with exercise. The N-terminal tyrosine residue in the endorphin sequences contains all the component elements for the phenyl-piperidine pharmacophore and can mimic the three-dimensional distribution of the nitrogen function relative to the aromatic ring which is required for receptor binding (Figure 2.19). Furthermore, adsorption of the peptide ligand onto the membrane will aid reorganisation of the amino acid sequence into the optimum and more rigid arrangement for binding to the receptor.

All the endorphins have some affinity for the μ receptor and, although this is an oversimplification, there is tendency for the enkephalins and the dynorphins to show higher propensities for the δ, and κ receptors whilst β-endorphin has high affinity for all three receptor subtypes. Opioid pharmacology is further complicated by the receptor subtypes having different distribution profiles within the body and that

TABLE 2.3
Amino Acid Sequences of Endogenous Morphine Peptides

Peptide	Peptide Sequence (Number of Amino Acid Residues)
Leu enkephalin	**Tyr–Gly–Gly–Phe–Leu** (5)
Met enkephalin	**Tyr–Gly–Gly–Phe–Met** (5)
Dynorphin A	**Tyr–Gly–Gly–Phe–Leu**–Arg–Arg–Ile–Arg–Pro–Lys–Leu–Lys–Trp–Asp–Asn–Gln (17)
Dynorphin B	**Tyr–Gly–Gly–Phe–Leu**–Arg–Arg–Gln–Phe–Lys–Val–Val–Thr (13)
β-endorphin	**Tyr–Gly–Gly–Phe–Met**–Thr–Ser–Glu–Lys–Ser–Gln–Thr–Pro–Leu–Val–Thr–Leu–Phe–Lys–Asn–Ala–Ile–Ile–Lys–Asn–Ala–Tyr–Lys–Lys–Gly–Glu (31)

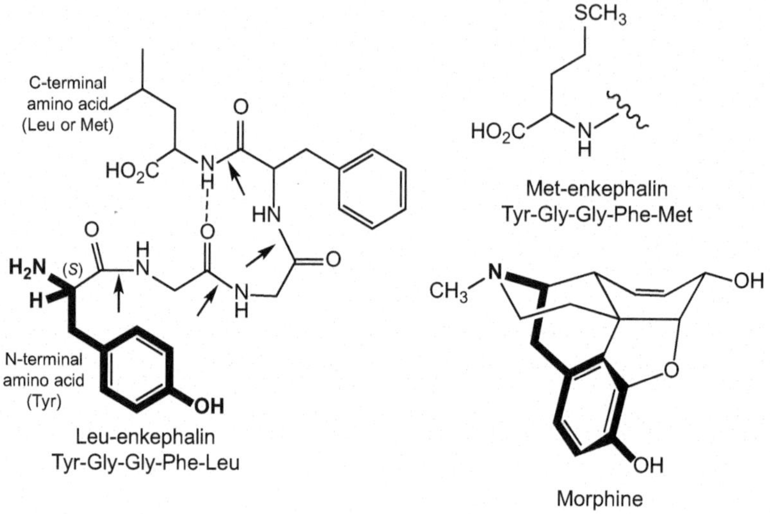

FIGURE 2.19 The two enkephalin pentapeptide sequences are shown on the left and top right. Most amino acids have the general structure $H_2N–CH(R_x)–CO_2H$ where R_x indicates the group which distinguishes one amino acid from another. Amino acids have (*S*)-stereochemistry at the chiral centre and form sequences of the type ~HN–CH(R_1)–CO–NH–CH(R_2)–CO~ when linked together in a peptide or protein sequence. The arrows in the Leu enkephalin diagram indicate the amide bonds (CO–NH) that form between the carboxylic acid group of one amino acid and the amino group of another amino acid in the peptide sequence. The key relationship between the N-terminal tyrosine residue of the enkephalin sequence and the morphine pharmacophore is highlighted in bold. See also Figure 2.30 for further details on amino acids and peptides/proteins.

stimulation of the various opioid receptors produces a range of side effects which are often dependent upon the location of the receptor, along with analgesia.

2.4.6 HOW MORPHINE WORKS AS AN ANALGESIC

Opioid receptors belong to a class called G protein-coupled receptors (GPCRs) which play an essential role in the communication between the external and internal environments of cells. GPCRs relay signals initiated by the binding of extracellular ligands such as hormones, neurotransmitters, ions, photons, drugs and other stimuli to internal secondary signalling systems which convey the message further, leading ultimately to the physiological response.

There are around 800 GPCR proteins and each one is highly specific to a particular initial signal. These proteins share a common structural motif in which seven transmembrane helices are connected to three extracellular and three intracellular loops but differ in the composition and length of the amino acid sequence. As the name implies, the GPCRs interact with G proteins which have the ability to bind to guanine nucleotides, GDP (guanosine diphosphate) and GTP (guanosine triphosphate). When morphine binds to its receptors the G protein stimulates conductance through potassium channels, decreases conductance through calcium channels and reduces cyclic adenosine monophosphate production via inhibition of the enzyme adenylate cyclase (Figure 2.20). Together these changes blunt the effect of signalling systems that transmit pain.

Pain is usually triggered by the activity of specialised damage-sensing neurons (nociceptors) which generate an action potential, leading to transmission of the initial signal from the peripheral nervous system to the spinal cord and brain. Ion channels play an important role at each step of the neuronal pain pathway; activation of sodium and calcium channels generates the action potential and potassium channels repolarise the membrane, inactivating sodium channels and returning the neuron to a resting state. Overexcitability of peripheral nociceptors is one of the major drivers of pain and the dysfunction of potassium channels as a result of administration of drugs such as morphine can limit the generation and firing rate of action potentials and alleviate the symptoms of pain.

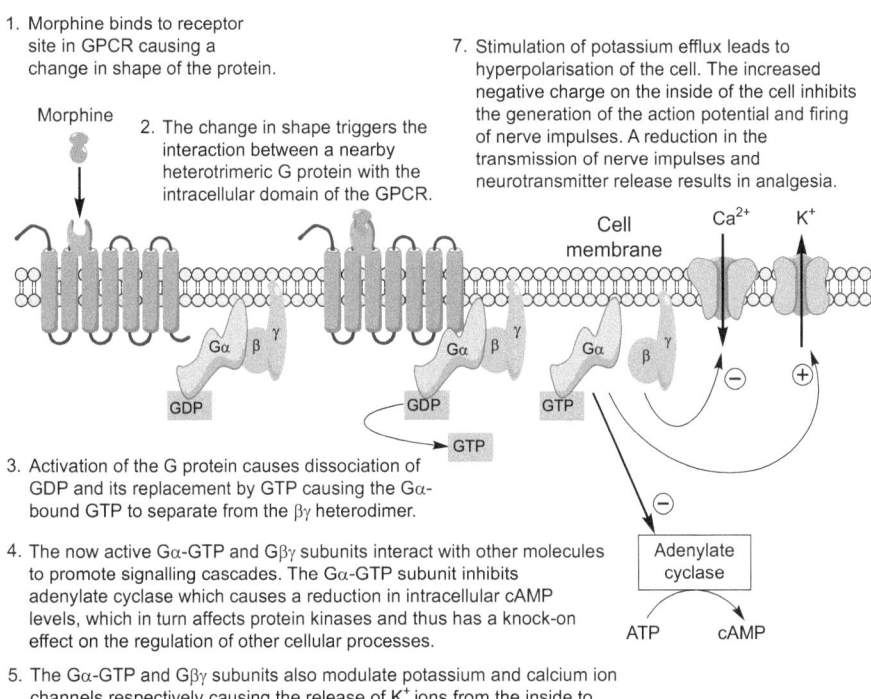

1. Morphine binds to receptor site in GPCR causing a change in shape of the protein.

2. The change in shape triggers the interaction between a nearby heterotrimeric G protein with the intracellular domain of the GPCR.

7. Stimulation of potassium efflux leads to hyperpolarisation of the cell. The increased negative charge on the inside of the cell inhibits the generation of the action potential and firing of nerve impulses. A reduction in the transmission of nerve impulses and neurotransmitter release results in analgesia.

3. Activation of the G protein causes dissociation of GDP and its replacement by GTP causing the Gα-bound GTP to separate from the βγ heterodimer.

4. The now active Gα-GTP and Gβγ subunits interact with other molecules to promote signalling cascades. The Gα-GTP subunit inhibits adenylate cyclase which causes a reduction in intracellular cAMP levels, which in turn affects protein kinases and thus has a knock-on effect on the regulation of other cellular processes.

5. The Gα-GTP and Gβγ subunits also modulate potassium and calcium ion channels respectively causing the release of K^+ ions from the inside to the outside of the cell and inhibiting the conductance of Ca^{2+} ions.

6. The Gα subunit will eventually hydrolyse the attached GTP to GDP by its inherent enzymatic activity, allowing it to reassociate with Gβγ, returning the system to resting inactive state.

FIGURE 2.20 A schematic outline of the sequence of intracellular events following the binding of morphine to a G protein-coupled opioid receptor that leads to the changes in ion channel conductance and inhibition of adenylate cyclase. Note that GDP is the abbreviation for guanosine diphosphate, GTP is for guanosine triphosphate, ATP is for adenosine monophosphate and cAMP is for cyclic adenosine monophosphate.

2.5   β-BLOCKERS AND CARDIAC DISEASE

2.5.1   ADRENERGIC RECEPTORS

β-Blockers act by selectively blocking adrenaline (Figure 2.21) from binding to its receptor and thus inhibit the cascade of cellular events which lead to a biological response. The adrenergic receptors are GPCRs and there are two types: α and β which are further classified into subtypes α_1, α_2, β_1 and β_2. Different types of G proteins are responsible for signalling through the different adrenergic receptor subfamilies. Upon adrenaline-mediated activation of β_2-adrenergic receptors for example, a stimulatory G protein increases the formation of cAMP (c.f. morphine receptor). The entire family of adrenergic receptors work in harmony to regulate the physiological functions of the sympathetic nervous system to produce the fight or flight response.

The β-blocker family of cardiac drugs was discovered through classic medicinal chemistry techniques before structural information on the receptor became available. However, it was known at the time of the discovery of β-blockers that adrenaline could bind to distinct receptor types within the cardiovascular and pulmonary systems. The main tissue locations of the different types of adrenergic receptors are outlined in Table 2.4 together with the physiological effect when the receptor is activated with adrenaline or noradrenaline. These changes prepare the body to handle the challenges of a fight or flight situation such as an increase in heart rate, blood flow and respiration and an energy boost from increased blood glucose due to the breakdown of glucagon. Adrenaline is used as an emergency treatment for cardiac arrest or severe anaphylactic shock but the wider clinical utility of adrenergic drugs lies in the selectivity which can obtained if a specific receptor is targeted. Salbutamol, for example, was designed as a selective β_2-stimulant bronchodilator therapy for asthma. Modulating activity of the β_1-receptor, specifically inhibiting the binding interactions which lead to elevated heart rates and blood pressure, was the basis for the design and development of the β-blockers.

2.5.2   AGONISTS AND ANTAGONISTS

Drugs that act by binding to receptors fall into two very broad categories: those that can bind to and stimulate the receptor to give the characteristic physiological response, which are known as agonists, or those that either block or, in some cases, reverse the action of the receptor, in which case they are termed antagonists (Figure 2.22). Salbutamol is an example of the former category whereas the β-blockers, by selectively blocking the endogenous ligand (adrenaline) at certain adrenergic receptors in

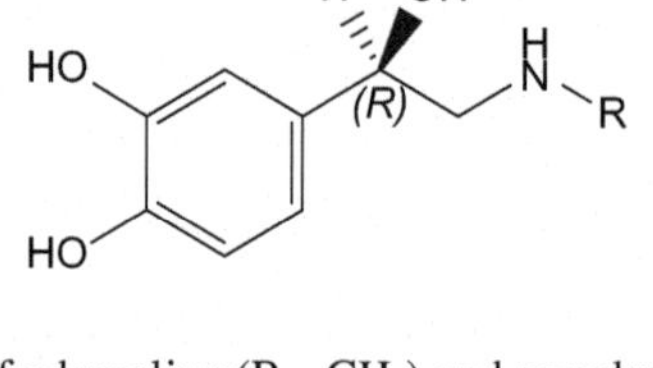

FIGURE 2.21   Structures of adrenaline (R=CH$_3$) and noradrenaline (R=H). Note that the chiral centre has the R-configuration.

TABLE 2.4
Distribution of Adrenergic Receptors in the Body and the Physiological Effects Resulting from Their Stimulation with an Endogenous Ligand

Receptor Subtype	Site of Action	Response to Stimulus
α_1	Peripheral blood vessels	Contraction of blood vessels ($\Rightarrow$ increased blood pressure and blood flow returning to heart).
α_2	Presynaptic nerve terminals	Regulates release of neurotransmitters; centrally induced sedation and analgesia.
	Pancreas	Inhibition of insulin secretion from pancreatic β cells via negative feedback loop.
β_1	Heart	Increases heart rate, increases the heart's strength of contraction.
β_2	Peripheral blood vessels	Vasodilation of peripheral vessels ($\Rightarrow$ increases blood flow to skeletal muscles).
	Bronchial smooth muscle	Bronchodilation ($\Rightarrow$ lets more air in and out during breathing).

Adrenaline and noradrenaline can both act as hormones and neurotransmitters. Adrenaline functions largely as a hormone and has slightly more of an effect on the heart whilst noradrenaline has more effect on blood vessels and is predominantly a neurotransmitter rather than a hormone. Both compounds are made and released by the adrenal glands.

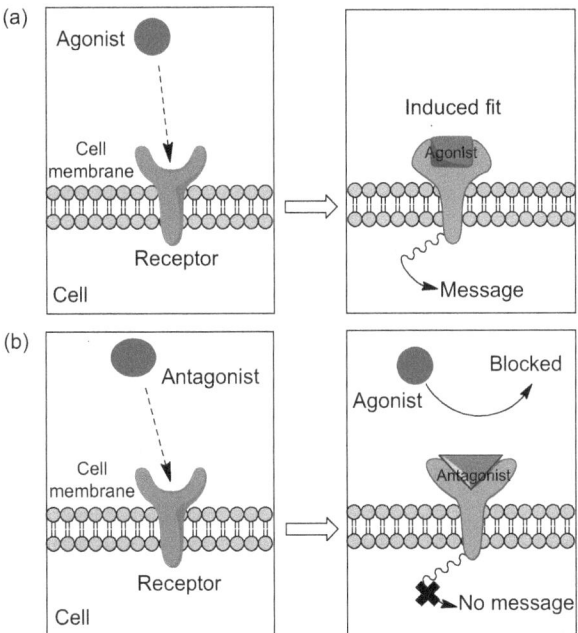

FIGURE 2.22 (a) Simple view of a receptor activated by an agonist and (b) how binding of an antagonist blocks the agonist from binding.

cardiac tissue, are antagonists. β-Blockers are predominantly used to treat a range of heart conditions such as high blood pressure, abnormal heart rhythms and angina but can also help manage anxiety and prevent migraines.

2.5.3   FIRST-GENERATION β-BLOCKERS: CARDIOSELECTIVE β-BLOCKERS

Several hundred compounds were synthesised and evaluated in the search for the elusive superior molecule which would block the β-receptor responsible for increased heart rate. Four key compounds involved in the development of propranolol, a compound which revolutionised the treatment of angina in the 1960s, are outlined in Figure 2.23. The driving force behind the design and development of propranolol at ICI Pharmaceuticals (Alderley Park, Cheshire) was Sir James Black, an academic pharmacologist who became an industrial pharmacologist. His pioneering work on drug discovery methodology and crucial role in the invention of two groundbreaking compounds, propranolol and cimetidine, a histamine H_2-receptor antagonist for the treatment of peptic ulcers, was recognised in 1988 with the award of the Nobel Prize in Medicine and Physiology.

The impetus for the β-blocker programme at ICI was the compound dichloroisoprenaline, which had evolved from the simpler adrenaline analogue, isoprenaline. The structural difference between adrenaline and noradrenaline is a methyl group on the nitrogen atom. Introducing an alkyl chain with a different number of methylene (CH_2) groups onto the nitrogen atom was a straightforward modification to the adrenaline structure and increased the size and lipophilicity of the compound which gave better membrane penetration. Isoprenaline, the N-isopropyl analogue, was a compound clinically used in the 1940s as a bronchodilator and cardiac stimulant, i.e. it was an agonist rather than an antagonist and acted almost exclusively on β-receptors.

Dichloroisoprenaline was the first β-blocker to be discovered and the pharmacological properties were reported in the literature in 1958. Although of no clinical significance, this compound was found to block the physiological responses of isoprenaline. This compound also exhibited stimulant actions, i.e. it was a partial agonist in that it had the ability to bind to and activate receptors in addition to a blocking effect. James Black translated these findings from an academic to an industrial context, the outcome of which was propranolol. Moreover, propranolol played an important role in acceptance of the dual receptor mechanism theory which up to that time had been viewed somewhat sceptically by influential members of the academic scientific community.

An early challenge in the β-blocker work was developing novel in vitro assays which would reveal the blocking properties of compounds that behaved as partial agonists in some systems. With suitable screening tests in place, the new synthetic compounds from the SAR studies could be readily assessed for inhibitory activity. One of the inspired structural modifications by the lead medicinal chemist on the team, John Stephenson, was replacing the two chlorine atoms of dichloroisoprenaline by another phenyl ring to make a naphthalene moiety. The naphthyl group would have the same spatial occupation as the dichloroaromatic moiety with the possible additional effect of extended π-bonding which might enhance the inhibitory activity. The naphthyl group with the extra aromatic ring could potentially result in a

Compound	Biological activity

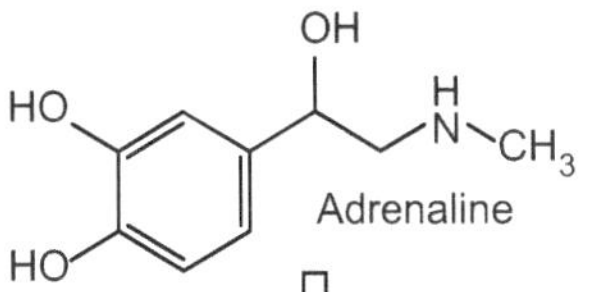

Adrenaline
- Both α-and β-receptor stimulant

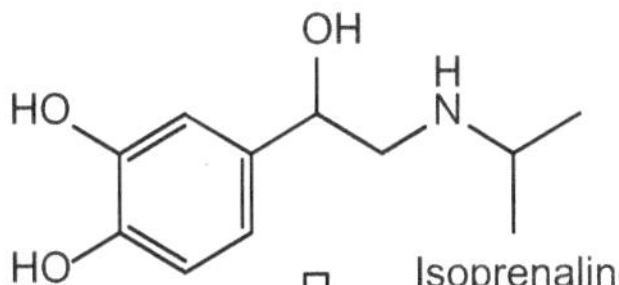

Isoprenaline
- Active for β-receptors over α-receptors **but** agonist not antagonist, no selectivity between different sub-types of β-receptor

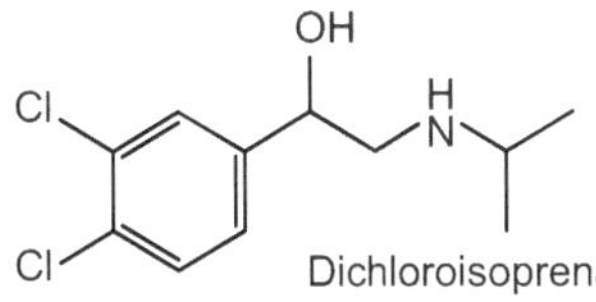

Dichloroisoprenaline
- 1st β-blocker but partial agonist
- Difference in activity could be due to increased lipophilicity (phenol is soluble in water whereas chlorobenzenes are not)

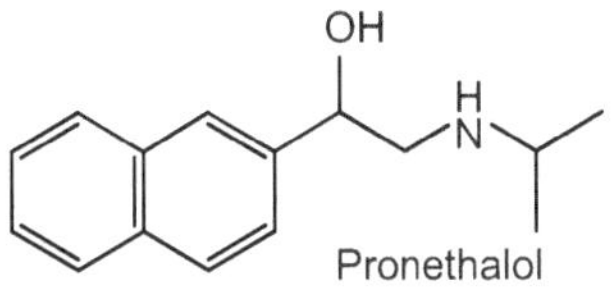

Pronethalol
- 1st β-blocker to be used clinically* to treat angina and had desirable effect of also reducing high blood pressure
- Innovative molecule and considered as 'first-in class' therapeutic
- Not in widespread clinical practice. Patented 1960, marketed 1963 and withdrawn soon after launch due to toxicity concerns

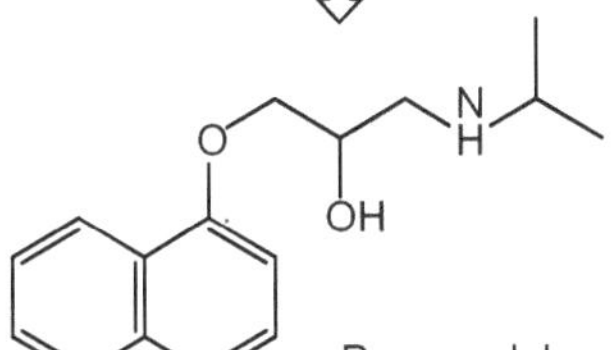

Propranolol
- Pure antagonist and initially used to treat angina
- 10-20 times more potent than pronethalol
- Approved for medical use 1964
- Used as bench mark against which all subsequently developed compounds were rated
- Subsequently found to be safe and effective treatment for managing a variety of conditions such as migraine and anxiety disorders in addition to a variety of cardiovascular conditions

FIGURE 2.23 Early β-blockers. *The transfer of pronethalol from lab (1960) to clinic (1963) took place before post-thalidomide regulations were introduced (Medicines Act 1968) when the pre-launch regulatory framework was more relaxed and less rigorous than the current practice.

hydrophobic interaction with the receptor in a part of the molecule which is not involved in binding. Such a molecule may bind to the receptor without activating the response. Pronethalol did indeed function as had been hoped; it was an antagonist with no sign of agonist activity in the laboratory and became the first β-blocker to be tested clinically and was effective in the treatment of angina.

Pronethalol was later withdrawn from clinical use due to toxic symptoms but a second and superior compound, propranolol, had been identified as part of the patent completion work. Propranolol, a compound 10 to 20 times more active and less toxic than pronethalol, emerged from an exploration into the effect of incorporating a bridging unit between the aromatic ring and the ethanolamine side chain. The synthesis of the compound with the oxymethylene (O–CH$_2$) group involved a base-promoted reaction of epichlorohydrin with naphthol followed by reaction with an amine to give the side chain moiety (Figure 2.24). Chance and luck, in addition to inspirational ideas, play a part in the drug discovery business. The chemist who first made propranolol could not find the bottle of β-naphthol as the starting phenol in the synthesis and so instead used α-naphthol, an isomeric compound with the OH group in a different position in the ring. The β-naphthyloxy analogue of propranolol in contrast turned out to be only slightly more potent than pronethalol and, of all the linking groups tried, the oxymethylene group proved to be the best. Pronethalol and propranolol have names with the suffix 'lol', which indicate that these compounds are β-blocking drugs although brand names of the particular pharmaceutical preparation may differ.

Propranolol was found to have widespread application for the management of a range of cardiovascular conditions such as hypertension, irregular heartbeat and heart failure. However, propranolol was a non-selective β-antagonist and acted at both the β_1 and β_2 receptors which caused problems of bronchoconstriction and airway obstruction in asthmatics. Propranolol use was also associated with the incidence, albeit low, of CNS side effects such as dizziness, vivid dreams and sedation and was attributed to the lipophilic nature of the compound which facilitated its transfer across the blood-brain barrier.

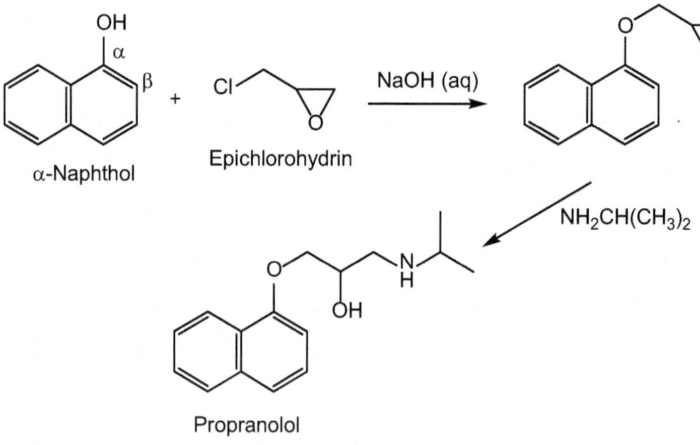

FIGURE 2.24   Synthesis of propranolol from α-naphthol. The two possible positions (α and β) for the OH group of naphthol are indicated.

Propranolol is usually administered as a racemic mixture but only (*S*)-propranolol exhibits the *β*-blocking activity in therapeutic doses whilst the *R*-enantiomer does not. The absolute configurations of the chiral centres in both adrenaline and propranolol are the same, i.e. the absolute arrangements of the substituents around the chiral carbon are the same. The *R* and *S* designatory prefixes are based on the Cahn-Ingold-Prelog rules which determine the priority of substituents and this changes between propranolol and adrenaline due to the additional oxygen atom linking the aromatic ring and side chain group.

2.5.4 SECOND-GENERATION *β*-BLOCKERS: SELECTIVE *β*₁-BLOCKERS

Propranolol is still considered to be the prototypical *β*-adrenergic receptor antagonist although there are concerns about its use in asthmatics. To address the selectivity issues extensive SAR studies were carried out, this time using propranolol as the lead compound. Quantifying structure-activity data was starting to emerge in the 1960s as a means of giving a more detailed understanding of how the physicochemical properties of compounds relate to their pharmacological activity, the results of which then aid the ensuing molecular design process. Quantitative SARs now form a sophisticated component of the modern drug discovery toolkit. The results of the investigation on propranolol analogues are summarised in Figure 2.25 and confirmed the earlier findings of the importance of the side chain group ($O–CH_2–CH(OH)–CH_2–NHR$), but there was scope for further changes to the aromatic ring system.

Cardioselectivity was found in a series of compounds which featured a *para*-amido substituent (NHCOR) on the aromatic ring (Figure 2.26). It was thought likely that the cardioselectivity arose from an additional hydrogen bond interaction between the *para*-substituent and a complementary site on the *β*₁- but not the *β*₂-receptor. Moreover, practolol and atenolol, which has an additional CH_2 group between the ring and the amide function, are also more hydrophilic than propranolol and achieve the goal of reducing CNS side effects. Atenolol, although not as potent as propranolol, is a safer and equally effective antihypertensive drug and was approved for medicinal use in 1975.

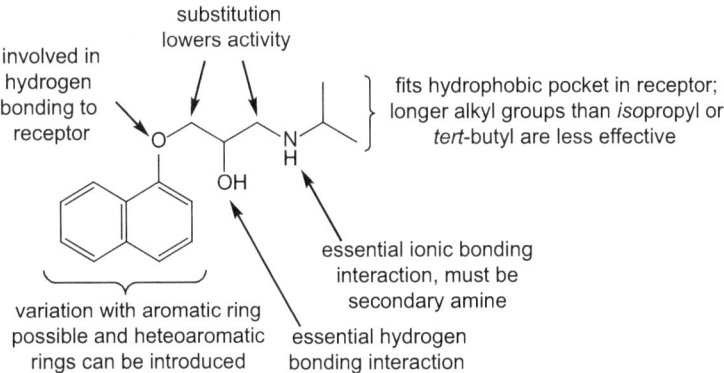

FIGURE 2.25 Structure-activity relationships on propranolol.

Compound Name (manufacturer) Relative potency (Propranolol = 1)	Comments
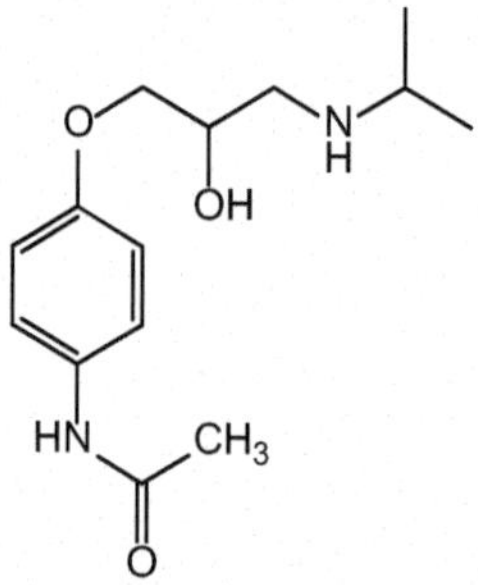 Practolol (ICI) 0.25	• Not as potent as propranolol but cardioselective, i.e. blocks cardiac β_1-receptors without blocking β_2-receptor. • Introduced in 1970 as 1st cardioselective β_1-blocker but withdrawn in 1975 because of side effects in small numbers of patients [skin rashes, eye problems and peritonitis (inflammation of peritoneum)].*

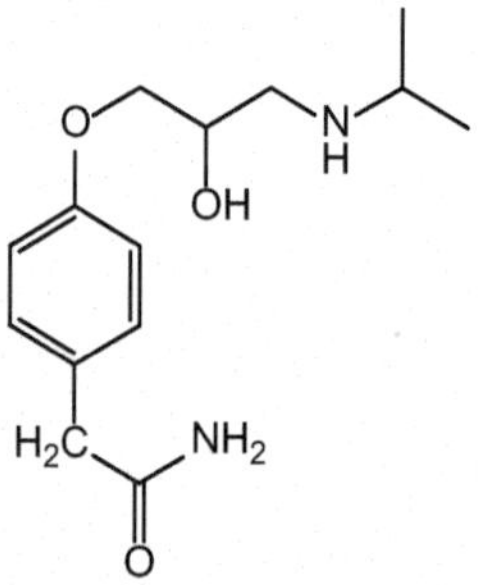 Atenolol (ICI)
0.80
	• The amide substituent in the aromatic ring must be in the *para*-position for selectivity for β_1-receptors.
	
	    • Selectivity due to additional hydrogen bonding interaction in β_1-receptors but not β_2-receptors
	    ⇒ Replacement of acetamido group with other groups capable of hydrogen bonding led to a series of cardioselective β_2-blockers ('me-too' medicinal molecules)
	
	• Used to treat hypertension, ischaemic heart disease, arrhythmias and migraine.

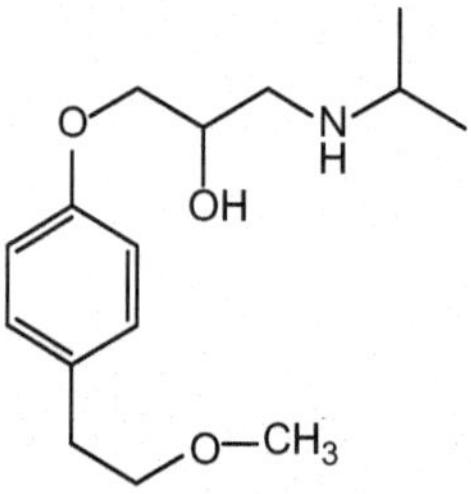

Metoprolol† (Hassle,
approved for use in 1974)
0.80

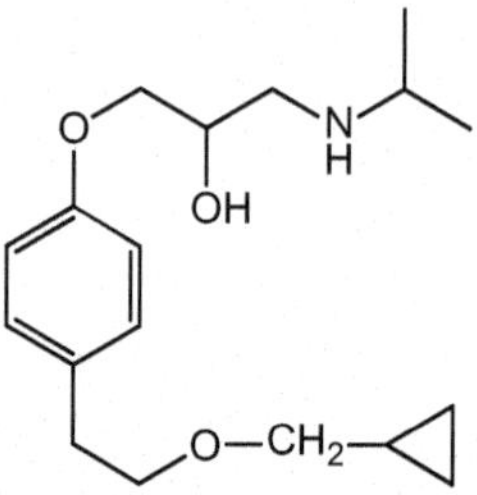

Betaxolol† (Synthelabo-Searle,
approved for use in 1983)
1.0

FIGURE 2.26 Cardioselective β_1-blockers. The drugs are administered as the racemate although the activity resides with the *S*-enantiomer of the compound. *Compensation scheme was set up by ICI as a result of the adverse side effects although the exact cause of some symptoms was never established. †Me-too drugs.

Pronethalol, propranolol, practolol and atenolol are a group of structurally related molecules that can be considered pioneering drugs which led and consolidated a particular drug company's lead (ICI) in a specific therapeutic area. Other companies were also working at the time on the aryloxypropanolamine area and, through demonstrating that their molecules had innovative design features, were able to patent and make their own versions of pronethalol, considered to be the 'first-in-class' β-blocker, and thus gain a share in a lucrative market. These compounds, sometimes called 'me-too' drugs, are not just for financial gain but serve a useful purpose such as offering improvements in activity, selectivity, side effects and drug delivery. Metoprolol is a short-acting β_1-selective compound and betaxolol has ophthalmic benefits and is used in topical preparations to treat glaucoma in which there is increased pressure in the eye.

2.5.5 THIRD-GENERATION β-BLOCKERS: HYBRID EFFECTS

In the first- and second-generation β-blockers with the general structure Ar–O–CH$_2$–CH(OH)–CH$_2$–NHR (where Ar indicates an aromatic ring system) the optimum groups on the nitrogen atom of the molecule were either $R = iso$propyl [–CH(CH$_3$)$_2$] or *tert*-butyl [–C(CH$_3$)$_3$]. Extension tactics involving addition of arylalkyl groups to the amine group resulted in compounds which were thought to bind to β_1-receptors through an extra binding interaction at this part of the molecule with the β_1-receptor. This modification, together with further changes to the aromatic ring (Ar) of the aryloxypropanolamine structure, led to a number of β-blockers with additional secondary beneficial effects. Carvedilol and nebivolol (Figure 2.27) are considered hybrid antihypertensive drugs which reduce blood pressure by two routes: inhibiting β_1-receptors and by a vasodilating action. Vasodilation is a product of several underlying mechanisms including blocking α_1-adrenergic receptors on peripheral blood vessels and inducing the release of endogenous nitric oxide (NO). The unique pharmacological profiles of the third-generation β-blockers afford certain advantages in the clinic, for example, in the treatment of diabetics with hypertension.

2.6 CAPTOPRIL, ENALAPRIL AND THE RENIN-ANGIOTENSIN SYSTEM

Blood pressure is regulated by a multitude of interrelated factors including not only the adrenergic neural system but also a number of peptide hormones. Some antihypertensive agents inhibit the action of enzymes in the renin-angiotensin system. A principal component of the renin-angiotensin system is the vasoconstricting octapeptide angiotensin II which is produced from the protein angiotensinogen by a series of hydrolysis reactions catalysed by the enzymes renin and ACE (Figure 2.28). Overproduction of angiotensin II can result in hypertension. Captopril and enalapril are ACE inhibitors and work by blocking the cleavage of the terminal His-Leu dipeptide from the C-terminus of angiotensin I.

Captopril is the pioneer drug (the breakthrough molecule for a new treatment of hypertension) whilst enalapril is a 'me-too' compound in which the basic structure of the molecule has been adapted to improve the pharmacological profile. The design of captopril is a further example of rational drug design, and the invention of this compound encompasses snake venom, key chemistry associated with enzyme inhibitor design, structural information on the binding site of ACE and inspiration (Figure 2.29). Enalapril is a prodrug and the ethyl ester group is hydrolysed in vivo to

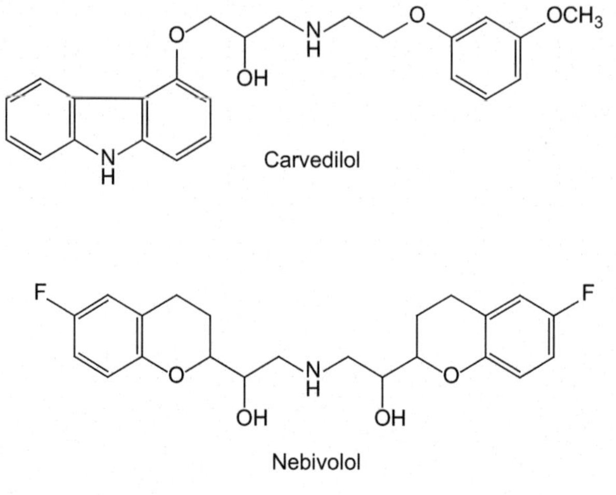

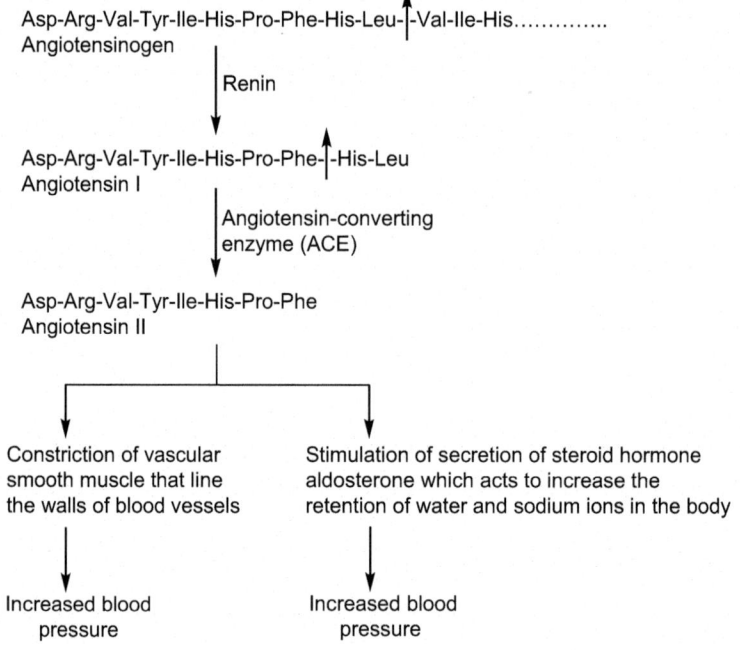

FIGURE 2.27   Structures of the third-generation β-blockers carvedilol and nebivolol.

Asp-Arg-Val-Tyr-Ile-His-Pro-Phe-His-Leu-|-Val-Ile-His..............
Angiotensinogen

Renin

Asp-Arg-Val-Tyr-Ile-His-Pro-Phe-|-His-Leu
Angiotensin I

Angiotensin-converting
enzyme (ACE)

Asp-Arg-Val-Tyr-Ile-His-Pro-Phe
Angiotensin II

| Constriction of vascular smooth muscle that line the walls of blood vessels | Stimulation of secretion of steroid hormone aldosterone which acts to increase the retention of water and sodium ions in the body |

Increased blood pressure | Increased blood pressure

FIGURE 2.28   The renin-angiotensin system. The three-letter abbreviations stand for individual amino acids. Renin catalyses the cleavage of the Val-Ile-His tripeptide from angiotensinogen, a 485-amino-acid protein to form angiotensin I. The C-terminal His-Leu dipeptide of angiotensin I is hydrolysed by ACE to give the octapeptide angiotenssin II. Two physiological actions of angiotensin II combine to have a powerful effect on blood pressure: contraction of smooth muscle tissue and increasing the volume of fluid in the vascular system through the actions of aldosterone.

pGlu-Trp-Pro-Arg-Pro-Gln-Ile-Pro-Pro

Teprotide

Benzylsuccinic acid

succinyl
component

proline
residue

The design of captopril was based on combining structural features from benzylsuccinic acid, a known inhibitor of carboxypeptidase A, with the C-terminal dipeptide Pro-Pro of teprotide, a nonapeptide which was an antihypertensive agent. As ACE was a dipeptidyl carboxypeptidase with a zinc ion at the active site which binds to a carboxylate group, it was reasoned that the two CO_2H groups in the prototype ACE inhibitor would need to be spaced further apart than in benzylsuccinic acid. It was therefore envisaged that the initial compound would be based on a succinyl derivative of proline.

(2R)-methylsuccinyl-(S)-proline

Exploration of a hydrophobic pocket in the receptor by substituting the succinyl moiety with methyl groups; the compound with a methyl group at the β-position was the most potent inhibitor. Four stereoisomers are possible in this molecule with two chiral centres and the most active inhibitor had the R-configuration at the methyl centre with the naturally occurring S stereochemistry for the proline residue.

Captopril

Replacement of the carboxyl group with the thiol (SH) group to give better binding to the Zn^{2+} ion in ACE.

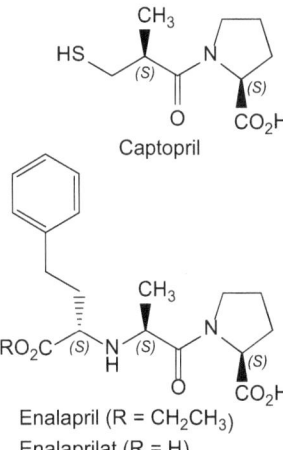

Enalapril (R = CH₂CH₃)
Enalaprilat (R = H)

Incorporation of a phenylethyl substituent next to the N-terminal carboxyl group to mimic a corresponding aromatic side chain in the known preference for ACE substrates to feature a hydrophobic amino acid in the third from last position. This would also compensate for the loss of interaction between the thiol group and the Zn^{2+} ion. The absolute stereochemistry of the methyl centre in captopril, enalapril and in (2R)-methylsuccinyl-(S)-proline is the same. The S designation in captopril and enalapril is based on the higher priority of the CH_2SH and NHR groups compared to the CH_2CO_2H group according to the Cahn-Ingold-Prelog rules.

FIGURE 2.29 Development of ACE inhibitors showing key compounds only. Many compounds were synthesised and tested during the SAR studies. Prior knowledge

(Continued)

FIGURE 2.29 (*Continued*)   leading up to the design of the synthetic inhibitors included the following. (a) Teprotide, a nonapeptide isolated from the venom of the pit viper *Bothrops jararaca*, was found to be a potent inhibitor of ACE and was the lead compound. (b) Work on peptidic inhibitors of ACE showed that they had a proline residue in the C-terminal position but peptides do not make good drug candidates as they are rapidly degraded by enzymes. (c) Recognition that carboxypeptidase A (Cpd A) was a surrogate model for the dipeptidyl carboxypeptidase ACE. Cpd A detaches one amino acid from the C-terminus of a peptide whilst ACE detaches two. (d) Benzylsuccinic acid was a specific inhibitor of Cpd A and resembled the two products from the enzymic reaction; the side chain and terminal carboxyl group from one of the amino acid products ($-CH(R)CO_2H$) joined to the carboxyl group of the second hydrolysis product from the reaction (a principle known as collected product inhibitor design). Thus the inhibitor binds to the active site of the enzyme and, as there is no scissile peptide bond present in this molecule, it stays attached to the enzyme therefore blocking its action. (e) Structural information on the Cpd A active site suggested that the aromatic ring of benzylsuccinic acid bound to a hydrophobic pocket, the C-terminal carboxylic acid (as the carboxylate anion) formed an ionic bond to a positively charged arginine residue and the N-terminal carboxylic acid (as the carboxylate anion) coordinated to the Zn^{2+} ion.

give the active dicarboxylic acid, enalaprilat. Captopril was approved for medicinal use in 1980 and enalapril four years later in 1984. Both compounds are chiral compounds and are marketed as single stereoisomers, and, as ACE inhibitors work by the same mechanism, they all end with the suffix 'pril'.

2.7   INSULIN, DIABETES AND rDNA

Insulin is a small protein hormone of 51 amino acids which plays a central role in the regulation of carbohydrates. A high concentration of blood glucose induces the secretion of insulin from the pancreas which promotes the uptake of glucose by tissues, particularly the liver and muscles. Blood glucose levels return to normal values and this in turn decreases the rate of insulin release. A deficiency in the body's ability to produce insulin leads to diabetes which, depending on the severity of the condition, may need to be managed by injections of insulin. Insulin has been available for the treatment of diabetes since the early 1920s, initially using purified animal pancreatic extracts and then in the 1980s with human insulin prepared by rDNA technology.

Insulin is composed of two chains of amino acids linked together by disulphide (-S–S-) bonds (Figure 2.30). Insulins from pigs and cattle are very similar in sequence and structure to that of the human protein which made it possible for animal insulin preparations to be used in the early years of diabetes therapy. Every protein, regardless of its source, is produced as a result of the expression of a specific genetic sequence coding for it. In rDNA technology, the DNA of a desired protein such as insulin is incorporated into a plasmid isolated from a bacterial cell and this modified plasmid is known as rDNA. This new plasmid is reinserted back into the bacteria, often *Escherichia coli*, which are then cultured and induced to express the desired protein (Figure 2.31).

The first recombinant human insulin produced involved chemically synthesising the genes for the A and B chains and forming two recombinant plasmids, one with the A-chain gene and one with B-chain gene, which were cultured separately in bacteria. Expression of the genes gave the two polypeptide chains which were

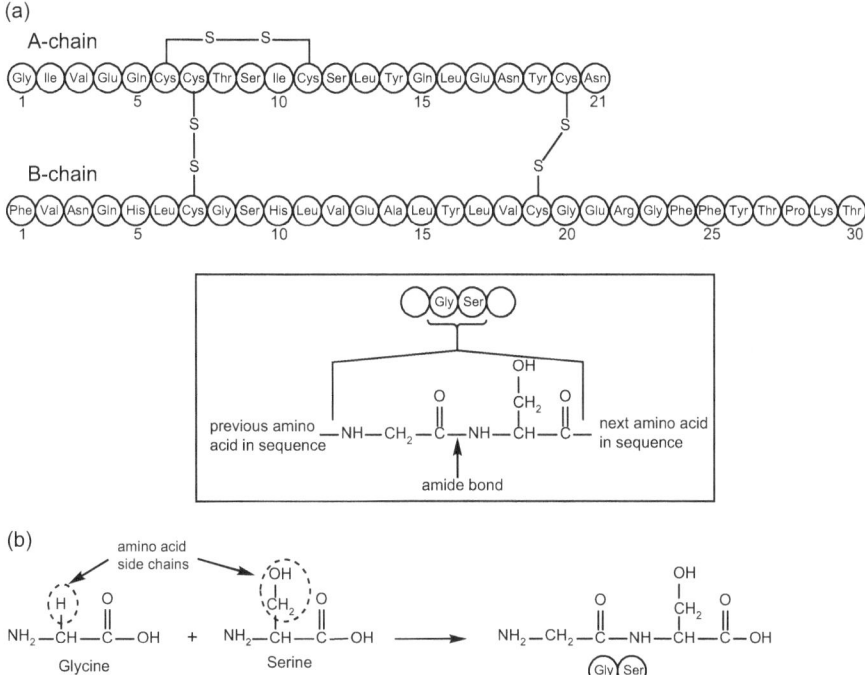

FIGURE 2.30 Human and animal insulins using three-letter codes for each of the amino acid residues, e.g. Gly is glycine and Ala is alanine. (a) Sequence of human insulin. Porcine insulin differs from human insulin by one amino acid; an alanine (Ala) is at position B30 rather than a threonine (Thr) residue. Bovine insulin differs in three positions compared to the human sequence: alanine (Ala) residues at A8 and B30 and a valine residue at A10. Before human rDNA was available porcine insulin was converted into the human sequence by replacing the terminal B30 alanine amino acid with a threonine residue using a combination of enzyme and synthetic chemistry methods. (b) The general formula for an amino acid is NH_2–CHR–CO_2H where R is the side chain. Amino acids differ from each other by the nature of their side chains. There are 20 amino acids found in proteins and the 20 side chains differ in respect of their structure, electrical charge and polarity and this gives each amino acid a unique chemical identity. Individual amino acids in a polypeptide or protein are linked together by amide bonds in which the carboxylic acid of one amino acid is specifically joined to the amine group of another amino acid. This is achieved biosynthetically with enzymes or synthetically using chemically modified amino acids to ensure the correct amide bond is formed.

isolated from their carrier proteins, purified and then linked chemically to form the disulphide bonds. Insulin is naturally synthesised in the pancreas as a single protein, proinsulin, in which the two chains are linked by a connecting sequence. This connecting peptide mediates the alignment of the molecule for the formation of the correct disulphide pairings and to fold into its three-dimensional shape. Enzymatic cleavage of the connecting peptide occurs in the storage granules of the Golgi apparatus of pancreatic cells to give the 51-residue insulin protein. The preferred commercial recombinant route to insulin now mimics this process as it requires only

1. The circular plasmid molecule is treated with restriction enzymes which recognise and cut specific sequences within the DNA.

3. The foreign gene fragment is enzymatically joined to the linearised plasmid to generate the circular recombinant plasmid, also known as an expression vector.

4. The recombinant plasmid is transferred to host cells such as the bacterium *E. coli* where they replicate and are induced to express the foreign protein that they would otherwise not produce.

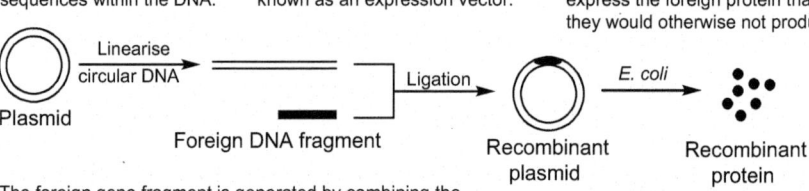

2. The foreign gene fragment is generated by combining the DNA sequence for the required protein (prepared by chemical synthesis or isolated from genomic DNA) along with a selection of codons which direct the host organism to synthesise the foreign protein. Restriction enzyme sequences are also incorporated at the ends of the foreign gene molecule to enable it to be inserted into the plasmid.

5. The foreign protein is isolated from cell extracts and purified, and where necessary processed further, to produce the fully functional protein.

FIGURE 2.31   Outline of the rDNA methodology for the production of recombinant proteins.

one genetically engineered plasmid encoding the proinsulin sequence and a single bacterial culture and purification protocol.

Insulin is stored within the pancreatic cells as hexameric units coordinated to two zinc ions. When insulin is secreted into the bloodstream, these zinc-insulin hexamers are diluted which causes the zinc ions to be released and the disassembly of the hexamers into the biologically active monomeric form of insulin. Normally, there is a feedback mechanism controlling insulin secretion from the pancreas based on changes in the blood glucose levels but there is no such feedback loop when insulin is administered to diabetic individuals and high or low blood sugar levels can occur if insulin injections are not timed correctly. The consequences of long-term exposure of tissues to high blood sugar levels cause the development of complications in diabetics such as heart disease, kidney problems and damage to the eyes and nervous system. Many insulin therapies have been designed over the years to optimise blood sugar control in diabetics and to more closely simulate the natural endogenous insulin response, i.e. a continuous low-level secretion with an additional release associated with meals. A standard therapy introduced in the mid-20th century designed to generate this pattern involved a mixture of a longer acting preparation in which insulin was complexed with zinc and other additives to delay absorption from the site of injection (usually subcutaneous tissue) into the bloodstream, together with monomeric insulin that had a faster onset of action.

Replacement therapy using more than one insulin formulation has continued to evolve but it is now possible to modify the amino acid sequence of insulin through genetic engineering and rDNA technology to develop analogues which are absorbed into the circulation at varying speeds. Insulin analogues have been produced which can either reduce or strengthen the interactions that hold the hexameric structures together, thus allowing for faster or slower disassembly and therefore the subsequent rate of monomer absorption into the bloodstream (Figure 2.32). Many clinically effective analogues incorporate changes to the insulin sequence at the end of the

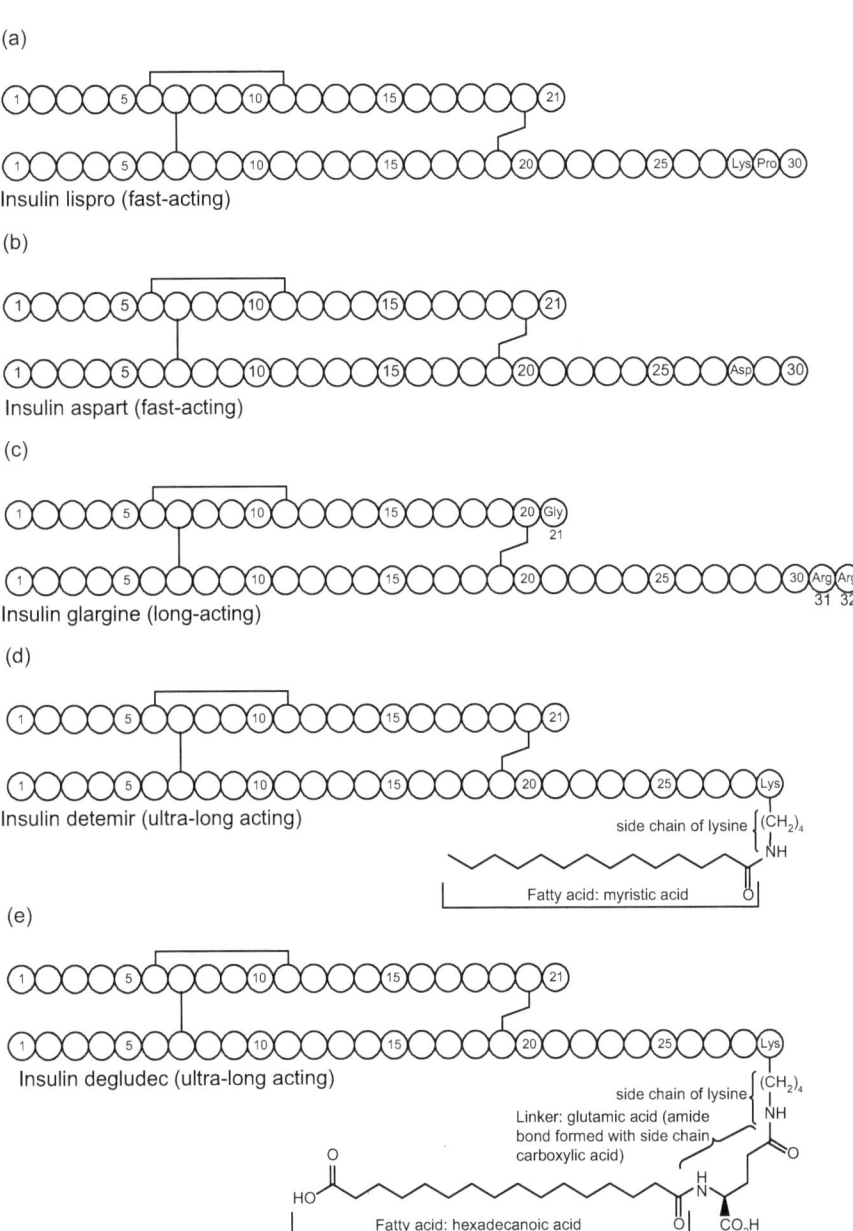

(a)

Insulin lispro (fast-acting)

(b)

Insulin aspart (fast-acting)

(c)

Insulin glargine (long-acting)

(d)

Insulin detemir (ultra-long acting)

side chain of lysine

Fatty acid: myristic acid

(e)

Insulin degludec (ultra-long acting)

Linker: glutamic acid (amide bond formed with side chain carboxylic acid)

side chain of lysine

Fatty acid: hexadecanoic acid

FIGURE 2.32 Structures of insulin analogues showing only the differences from the natural human insulin sequence. (a) and (b) Faster-acting analogues. The design of insulin analogues lispro and aspart was based on knowledge of the crystal structure of insulin and identifying residues which had a role in the association interactions between monomer units and the likely consequent effect on hexamer stability if these interactions were weakened. (c) Insulin glargine. The development of this compound was centred on altering the overall charge on the molecule. When the net charge of the amino acid side chains in a protein

Figure 2.32 (*Continued*)   is zero (known as the pI) the molecule has low solubility in a solution with the corresponding pH. Subcutaneous tissue has a neutral pH and the addition of basic arginine residues raised the pI towards a neutral pH value. (d) and (e) Fatty acid acylated insulins. The fatty acid substituents are attached to the side chain ε-amino group of the B29 lysine residue, and in the insulin degludec molecule, the attachment is via a linker group. These insulin analogues bind to albumin through non-polar and ionic interactions and remain in circulation longer as albumin has a slow rate of clearance from the blood.

B-chain as structure-activity studies have shown that the last three amino acids in the B-chain can be removed without loss of activity and that the B26-B30 region of the molecule is not critical in binding to the insulin receptor. Moreover, porcine insulin differs from the human sequence by only one amino acid, at the terminal position of the B-chain, and has been used safely in diabetes treatment for many years. Reversing the normal sequence of proline at position 28 and lysine at position 29 at the terminus of the B-chain of human insulin into the Lys^{28}-Pro^{29} sequence or replacing the proline with an aspartic acid residue decreased the stability of the hexamers and generated the rapid-acting analogues insulin lispro and insulin aspart. These analogues take effect quicker than regular human insulin which allows for administration much closer to meal times and is useful if a diabetic person wants to have greater flexibility in choosing their meal times. Insulin glargine in contrast is a slow-onset and long-acting insulin analogue and differs from the native sequence by the addition of two arginine residues at the end of the B-chain and terminating the A-chain with a glycine rather than an asparagine amino acid. Whilst this compound is completely soluble under the acidic conditions of the formulation, it precipitates and forms higher order aggregates of insulin hexamers at physiological pH, resulting in a slow dissolution process to give a relatively constant concentration of insulin over a 24-hour period.

Insulin with an ultra-long duration of action has been achieved through covalently attaching a long fatty acid moiety to the lysine at position 29 of the B-chain and omitting the terminal B30 threonine residue. The prolonged duration of action of insulin detemir and insulin degludec is thought to be mediated by two mechanisms: multi-hexamer structure formation at the site of injection and subsequent binding via the fatty acid chain of the monomers to albumin in the bloodstream.

The availability of insulin analogues with a range of pharmacokinetic profiles offers increased flexibility to the treatment regimens on offer to people with diabetes. Blood glucose monitoring devices and insulin pen delivery systems are also helping diabetics achieve greater glycaemic control and improve their quality of life.

2.8   CONCLUDING REMARKS

The design of a new medicinal molecule needs to start with something – be that folklore in the case of aspirin and opioid analgesics, a naturally occurring endogenous substance (adrenaline and the β-blockers) or identifying a protein deficiency which causes the disease (insulin and diabetes). Rounds of experiments then follow to improve on the starting compound and, ultimately, perhaps make a drug which is a safe and effective treatment for the target condition. This has traditionally been the realm of the medicinal chemist; exploring how small molecular changes can impact a compound's biological activity and toxicity profile, but this traditional role is expanding with the development of biologics. Human insulin was the first of the biologic therapeutics; compounds such as recombinant proteins, monoclonal antibodies and vaccines (including those for Covid-19) produced by living cells grown in vitro, i.e. in a laboratory setting. Insulin degludec, however, is a hybrid structure prepared by covalently binding a small molecular component onto a modified insulin sequence. This approach whereby small synthetic molecules are combined with larger biologic molecules is opening up new avenues for innovative drug design.

Monoclonal antibodies are the biggest group of biologic therapeutics and have a significant role in the treatment of a range of diseases including cancer. Cytotoxic molecules can also be bonded to tumour-targeting monoclonal antibodies and the 'magic-bullet' concept, proposed by Paul Ehrlich in the early 20th century, is becoming a reality in cancer treatments. These antibody-drug conjugates are comprised of an antibody (to target the cancer cell without affecting healthy cells) linked to a cytotoxic drug (the 'payload') through a chemical linker. The linker is designed to maintain the conjugate in an inactive non-toxic state whilst circulating in the blood and then release the cytotoxic agent through a chemical reaction once the conjugate becomes internalised within the tumour cell. In effect, these compounds are pro-drugs in which medicinal chemists have built into the structures certain design features which will only permit the therapeutic potential of the molecule to be realised once the correct reaction conditions have been met, i.e. within the cell itself.

The β-blockers and opioid analgesics were developed at a time when structural biology was in its infancy but they are now known to exert their activity through binding to transmembrane GPCRs. Detailed structural information on these proteins, and their interaction with known agonists and antagonists, is now emerging through X-ray crystallography and more recently cryo-electron microscopy (cryo-EM). X-ray crystallography and nuclear magnetic resonance (NMR) techniques are well-embedded into the design phase (Figure 2.10) of medicinal molecule development and give structural information at the atomic level for compounds which can be crystallised or dissolved in solution. Cryo-EM is revolutionising the understanding of the structure of membrane proteins, macromolecules which are not readily crystallisable or soluble, but have not yet reached the level of detail obtained from X-ray and NMR methods. However, cryo-EM has the potential to give deeper insight into the size, three-dimensional shape and amino acid composition of the binding pockets of these and other important protein drug targets and could drive the design process in new directions. One possibility is that with knowledge of a three-dimensional template for a druggable space, computational and medicinal chemists could design and synthesise compounds which have all the required binding characteristics to interact with the target molecule, but with completely unique and unexpected structures.

Natural products and compound library screening continue to give new lead and clinical candidate molecules. Paclitaxel (see Figure 2.1, and also known as taxol), an anticancer drug, has a complex tetracyclic structure with 11 chiral centres and was originally derived from the bark of the Pacific yew tree, *Taxus brevifolia*. The bark of a tree is not a sustainable source for a medicinal molecule, but a compound with the core tetracyclic ring moiety of paclitaxel was found in the leaves of the European yew tree, *Taxus baccata*. Medicinal chemists were then able to convert this core structure, harvested from the sustainable source, into paclitaxel in a few synthetic steps. Azidothymidine (AZT), the first of the drugs used to treat HIV/AIDS, was identified through the screening of a pre-existing compound library. More recently a novel class of macrocyclic peptide with the potential to tackle some of the most highly antibiotic-resistant infections has been identified through a screening programme. One of these compounds, zosurabalpin, is in clinical trials but is still a long way from being a clinically approved drug.

The origins of aspirin and morphine are rooted in folklore and can be considered serendipitous discoveries. Curiosity and perseverance of scientists have also given, for example, the penicillin antibiotics and the anticancer drug cisplatin and may lead to more novel therapeutic compounds in the future. In the late 1980s, a group of Belgian research scientists stumbled across smaller, simpler versions of normal antibodies in the immune system of camels. These mini-antibodies, or nanobodies, are single-chain protein molecules around half the size of the normal antibodies and are produced in response to infection in not only camelids but also in sharks. The small size of these compounds enables them to reach and bind to parts of target molecules that normal antibodies cannot reach and thus they have enormous therapeutic potential for diseases including autoimmune disorders, cancer and viral infections such as coronavirus. In 2018, the world's first nanobody-based drug, caplacizumab, was approved for the treatment of acquired thrombotic thrombocytopenic purpura, a rare blood-clotting disorder, and other nanobody drug candidates are in various phases of clinical trials.

FURTHER READING

The topics covered in this chapter are discussed in greater depth in the books and academic articles listed below and original material can be sourced from the bibliographies contained within. Medicinal molecules continue to be a developing and expanding field and a number of research papers on recent material are included.

BOOKS

Bladon, C. M. (2002) *Pharmaceutical Chemistry: Therapeutic Aspects of Biomacromolecules*. Chichester: Wiley.

Emery, A. E. H. and Malcolm, S. (1995) *An Introduction to Recombinant DNA in Medicine*. 2nd Ed. Chichester: Wiley.

Ganellin, C. R. and Roberts, S. M. (eds.) (1993) *Medicinal Chemistry: The Role of Organic Chemistry in Drug Research*. 2nd Ed. London: Academic Press.

Patrick, G. L. (2023) *An Introduction to Medicinal Chemistry*. 7th Ed. Oxford: Oxford University Press.

Wermuth, C. G., Aldous, D., Raboisson, P. and Rognan, D. (eds.) (2015) *The Practice of Medicinal Chemistry*. 4th Ed. London: Academic Press.

ARTICLES

Congreve, M., de Graaf, C., Swain, N. A. and Tate, C. G. (2020) Impact of GPCR Structures on Drug Discovery. *Cell*. [Online] 181 (1). pp. 81–91. Available from: https://doi.org/10.1016/j.cell.2020.03.003. [Accessed: 17th January 2024].

Fu, Z., Li, S., Han, S., Shi, C. and Zhang, Y. (2022) Antibody Drug Conjugate: The 'Biological Missile' for Targeted Cancer Therapy. *Signal Transduction and Targeted Therapy.* [Online] 7. Article no. 93. Available from: https://doi.org/10.1038/s41392-022-00947-7. [Accessed 17th January 2024].

Hirsch, I. B., Juneja, R., Beals, J. M., Antalis, C. J. and Wright, Jr., E. E. (2020) The Evolution of Insulin and How it Informs Therapy and Treatment Choices. *Endocrine Reviews.* [Online] 41 (5). pp. 733–755. Available from: https://doi.org/10.1210/endrev/bnaa015. [Accessed 29th August 2023].

Makurvet, F. D. (2021) Biologics vs Small Molecules: Drug Costs and Patient Access. *Medicine in Drug Discovery.* [Online] 9 (March). Article no. 100075. Available from: https://doi.org/10.1016/j.medidd.2020.100075. [Accessed 16th January 2024].

Odoemelam, C. S., Percival, B., Wallis, H., Chang, M.-W., Ahmad, Z., Scholey, D., Burton, E., Williams, I. H., Kamerlin, C. L. and Wilson, P. B. (2020) G-Protein Coupled Receptors: Structure and Function in Drug Discovery. *RSC Advances.* [Online] 10. pp. 36337–36348. Available from: https://doi.org/10.1039/D0RA08003A. [Accessed 16th September 2023].

Piper, S. J., Johnson, R. M., Wootten, D. and Sexton, P. M. (2022) Membranes under the Magnetic Lens: A Dive into the Diverse World of Membrane Protein Structures Using Cryo-EM. *Chemical Reviews.* [Online] 122 (17). pp. 13989–14017. Available from: https://doi.org/10.1021/acs.chemrev.1c00837. [Accessed 16th January 2024].

Zampaloni, C., Mattei, P., Bleicher, K., Winther, L., Thäte, C., Bucher, C., Adam, J.-M., Alanine, A., Amrein, K. E., Baidin, V., Bieniossek, C., Bissantz, C., Boess, F., Cantrill, C., Clairfeuille, T., Dey, F., Di Giorgio, P., du Castel, P., Dylus, D., Dzygiel, P., Felici, A., García-Alcalde, F., Haldimann, A., Leipner, M., Leyn, S., Louvel, S., Misson, P., Osterman, A., Pahil, K., Rigo, S., Schäublin, A., Scharf, S., Schmitz, P., Stoll, T., Trauner, A., Zoffmann, S., Kahne, D., Young, J. A. T., Lobritz, M. A. and Bradley, K. A. (2024) A Novel Antibiotic Class Targeting the Lipopolysaccharide Transporter. *Nature.* [Online] 625. pp. 566–571. Available from: https://doi.org/10.1038/s41586-023-06873-0. [Accessed 17th January 2024].

WEBSITE

World Health Organisation. (2023) *Model List of Essential Medicines.* [Online] Available from: https://list.essentialmeds.org/. [Accessed: 20th January 2024].

3 Riches of the Rocks

3.1 INTRODUCTION

3.1.1 FROM GEOLOGY TO ARCHAEOLOGY

A glance at the geological map of Britain and Ireland, Figure 3.1a, shows a complex pattern of shapes and colour which represent the different rocks and indicates an eventful geological past for such a relatively small land mass. Making sense of this jigsaw puzzle is complicated but some clear underlying trends can be detected if the different types of rocks and their chronological age are considered.

First consider the maps indicating metamorphic and sedimentary rocks, Figure 3.1b and c. There is a striking difference between the types and ages of the rocks found in northern Scotland and other parts of the British Isles; rocks in England are predominantly sedimentary and young compared to those in northern Scotland which are much older and metamorphic. This difference is because the two areas were once part of separate land masses which collided and joined during the Caledonian orogenesis, a tectonic process which took place roughly between the late Cambrian and mid-Devonian periods about 490 and 390 million years ago. The oldest rocks in Scotland are Lewisian gneiss, dating back around 2.6 billion years, and make up the Outer Hebrides, parts of the Inner Hebrides and the NW Scottish mainland.

Secondly, on the map of metamorphic rocks there is a recognisable boundary running from Stonehaven near Aberdeen in the north-east of Scotland to the island of Arran in the south-west which can be further extended into Ireland. This line is the Highland Boundary Fault and it separates the Highlands from the Lowlands. The fault line indicates the period of mountain building, the Caledonian orogeny, caused by the closure of the ancient Iapetus Ocean and the collision of continents. During this major geological event the existing rocks were folded and uplifted to form high mountain ranges and were subjected to extensive metamorphosis in the process. The Great Glen and Southern Upland faults also show this NE-SW trend (Inverness to Fort William and Dunbar to Girvan respectively) which is further evidence for the direction of the collision between plates in the Caledonian orogeny. The near vertical fault running from Cape Wrath on the north-west coast of Scotland to Skye is known as the Moine Thrust and represents an earlier period of mountain building in the Pre-Cambrian era.

A third trend is that the youngest of the sedimentary rocks, Figure 3.1c, are in the south-east around London and progress in age in a north-westerly direction. Finally, igneous rocks, Figure 3.1d, are found scattered across Scotland, Ireland, NW Wales and the south-western and northern parts of England. Some of these rocks, particularly granites in Scotland and northern England, were formed during the Caledonian orogenesis when there was extensive volcanic activity.

Geology and geological processes shape landscapes. For example, the high-level plateaux of the Cairngorms in the Highlands of Scotland are granite, a hard and

DOI: 10.1201/9781003562313-3

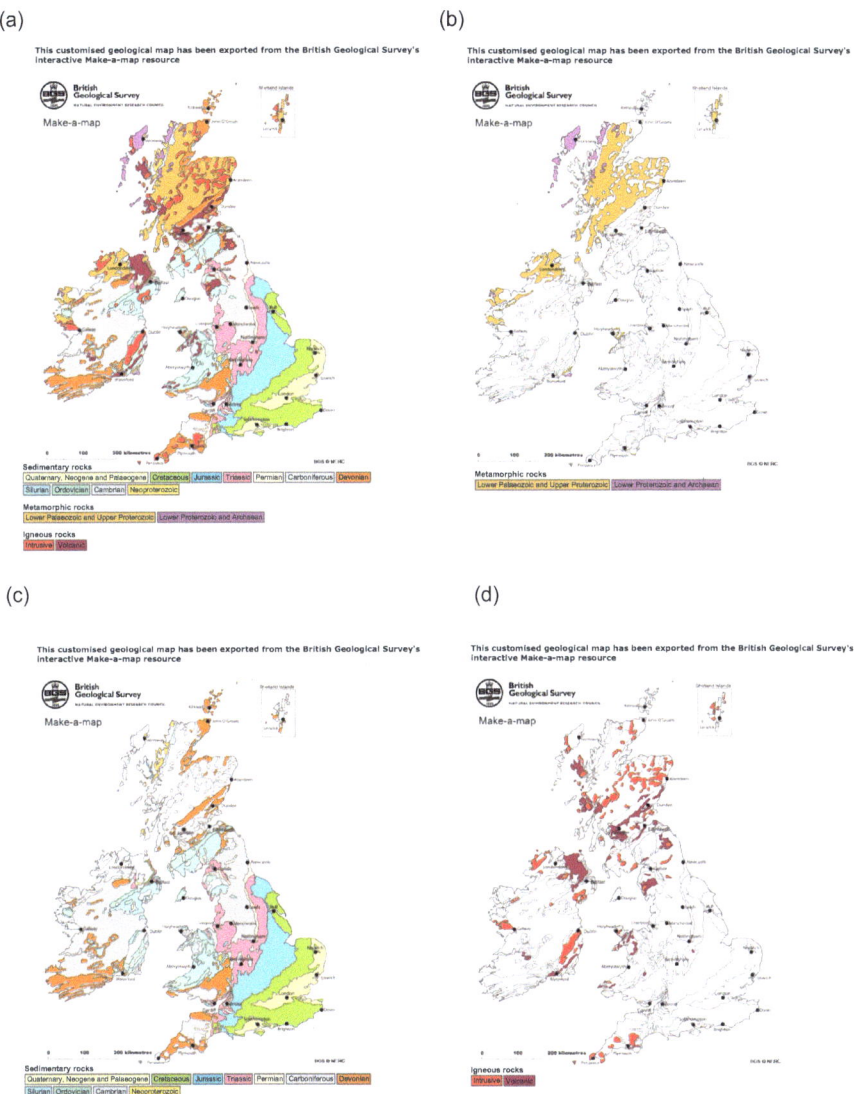

FIGURE 3.1 Outline geological maps contain British Geological Survey materials © UKRI [2024] from BGS Make-a Map (https://www2.bgs.ac.uk/discoveringgeology/geologyofbrit-ain/makeamap/map.html). These maps show the unweathered rock below the soil surface and each different rock type is shown by a colour. (a) Complete geological map of the Britain and Ireland, (b) metamorphic rocks, (c) sedimentary rocks and (d) igneous rocks.

impermeable igneous rock which has weathered over millions of years to form a rounded and bare sub-arctic mountain landscape. This landscape covers approximately 300 square miles and includes the largest area of land over 900 m and has five of the six highest mountains in the UK. The granite domes formed as part of the Caledonian orogenesis when molten rock (magma) pushed its way up into overlying metamorphosed rock. As the intrusions of molten magma cooled to form granite, some areas were subjected to further alteration by hot fluids moving through fracture zones causing softer and more crumbly rock. The altered fractured rock eroded more quickly to form valleys and then glacial erosion during the last Ice Age carved further into the rocks to produce more varied landforms of corries, deeply eroded glens and wider river valleys.

Although the Cairngorms area is known for its large-scale natural wilderness and remoteness, the influence of humans is in evidence in the farmed glens and valleys and in the infrastructure associated with outdoor recreational activities. Historically, the relative isolation of the area, harsher climates and limited scope for agriculture resulted in Highland communities developing their own unique cultural identity: speaking Gaelic with a social structure based on the clan system. A degree of animosity existed between the communities in the Highland and Lowland regions which became heightened in the 18th and 19th centuries as political changes and the dismantling of traditional ways of life were imposed from the south.

Aberdeen on the coast of Scotland is sometimes known as the 'granite city' as many of its civic and notable buildings were constructed from granite cut from local quarries. Granite is an attractive stone for buildings as it is hard and durable and mineral crystals sparkle in the sun. However, it is expensive and often is now only used for the facades of 21st-century buildings made with concrete and steel core structures. The traditional building material of London is brick, manufactured from local clay. London clay is young in geological terms, being deposited around 55 to 34 million years ago in the Eocene period, on top of an older layer of Cretaceous chalk which had been folded to form a basin. The chalk beneath London comes to the surface in the hills of the Chilterns to the North and the North Downs to the south with the latter terminating at the eastern end by the White Cliffs of Dover. The soft London clay is moreover easy to tunnel through which enabled the creation of the capital's extensive underground railway system. Whilst bricks produced from London clay built many of the capital's residential streets, more iconic architectural structures such as the British Museum, St Paul's Cathedral and government buildings of Whitehall used imported stone such as limestone from Portland in Dorset.

Some buildings have come to represent more than just an architectural construction. Big Ben, the tower and clock, at the north end of the Houses of Parliament in London, for example, is not only a prominent landmark in the capital but is also a symbol of UK democracy. Another iconic monument is Stonehenge, the lintelled stone circle in Wiltshire built and rebuilt between 3,000 and 1,600 BCE. Built as part of a larger cultural complex, the ruin which stands today has been studied extensively by archaeologists and symbolises the creativity, ingenuity and technological skills of the Neolithic people.

The cultural dimension of rocks is explored in this chapter through a sequence of topics ranging from what makes one rock different from another to geology-related aspects of archaeology. The chapter starts with the essentials of what constitutes the common igneous, sedimentary and metamorphic rocks (Section 3.2). Observing and handling geological samples helps to identify the different types of rock present in rural and urban landscapes and suggestions are given for this aspect of practical geology. Minerals, components particularly of igneous rocks, have specific chemical formulae and the external shape of these compounds reflects the internal arrangement of the ions. Section 3.3 first takes a closer look at the carbonate minerals calcite and aragonite, which both have the formula $CaCO_3$, and then silicate minerals with a particular emphasis on those of granite. Minerals are also considered from a mining perspective in Section 3.4 with an overview of the extraction and processing of metal ores from the copper age to the 21st century. A distinctive feature of sedimentary rocks is the presence of fossils, the preserved remains of ancient plants and animals, which give a tangible link to life in the past. Section 3.5 considers the fossilisation process itself and the significance of fossils in the sequencing of rock strata in the 19th century which subsequently led to the geological timeline (see below). Numerical dates shown in the geological timeline were obtained in the 20th century through analysis of the decay of radioactive isotopes in minerals and the basis of these techniques is the subject of Section 3.6. Decay of isotopes with half-lives shorter than those used to date rocks gives data relevant to an archaeological span of time. Analysis of archaeological human remains gives insight into the lives of these individuals and examples of this type of study are illustrated in Section 3.7. Stonehenge continues to fascinate and geoarchaeological studies identifying the geographical location of the sarsen and bluestones reveal yet more detail on Neolithic communities and a summary of these findings concludes the chapter.

3.1.2 GEOLOGICAL AND ARCHAEOLOGICAL TIME

Geological history covers a very long period of time but can be divided into eras (Palaeozoic, Mesozoic and Caenozoic) according to the different fossils present in rock strata. These three eras of time are then further split into periods, e.g. the Tertiary and Quaternary together form the Caenozoic era (Figure 3.2). The Quaternary period covers the last 2.6 million years and is sometimes referred to as archaeological time as it starts at the point when the first recognisable artefacts, items made or used by humans, such as stone tools appeared. The end of the last Ice Age and a warmer and drier climate around 12,000 years ago represents the start of the current sub-division of geological time, known as the Holocene epoch and in archaeological terms this was the beginning of the Mesolithic age.

3.2 IGNEOUS, SEDIMENTARY AND METAMORPHIC ROCKS

Rocks can broadly be considered to be aggregates of mineral particles which are inorganic compounds with a distinct chemical composition. Some rocks such as chalk consist almost entirely of one mineral – nearly pure calcium carbonate ($CaCO_3$) whilst others contain several. Granite, for example, is largely composed of quartz (SiO_2), potassium (or alkali) feldspars (such as $KAlSi_3O_8$) and mica $K(Mg,Fe)_3AlSi_3O_{10}(OH,F)_2$.[1] Classification and identification of different rocks into one of three groups: igneous, sedimentary or metamorphic, depends on the type and amount of minerals present together with textural appearance and structural features. For example, chalk is a fine-grained pale coloured sedimentary rock whilst igneous granite contains larger and shiny interlocking crystals with well-defined shapes. Mineral composition however can vary widely from one rock to another which complicates the identification of

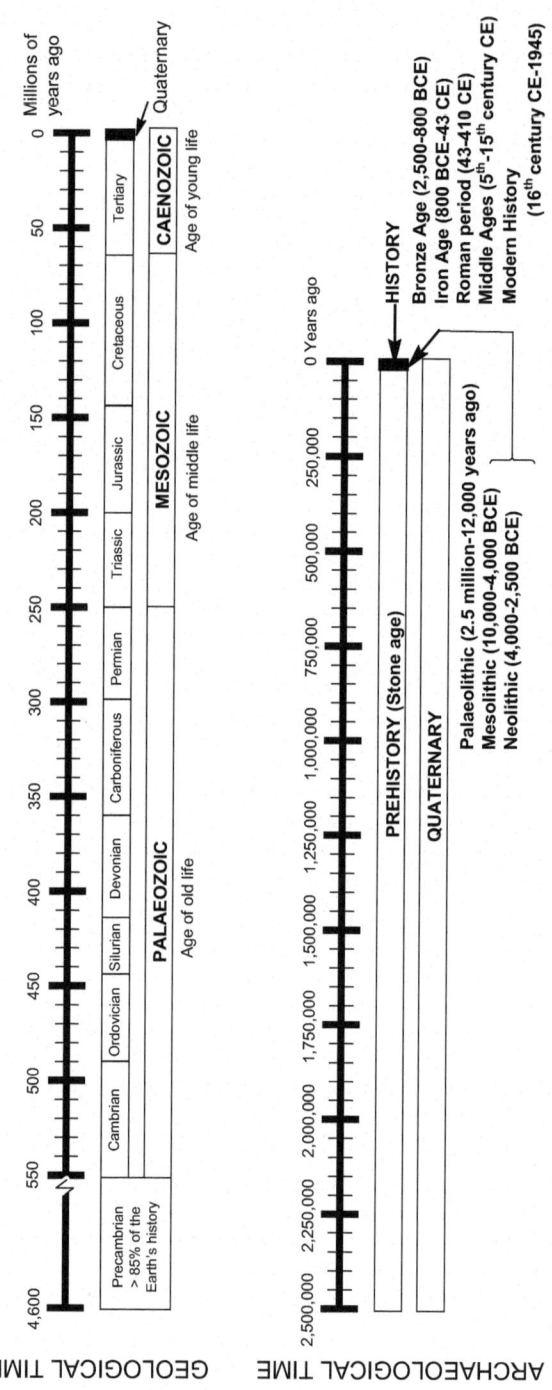

FIGURE 3.2 Geological and archaeological timelines. The archaeological timeline spans the time of human existence and covers the Quaternary period which started approximately

(Continued)

FIGURE 3.2 (*Continued*)   2.6 million years ago and the beginning of human tool use. Some archaeologists use an extended timeline starting at around five to six million years ago which marks the time when the human lineage split from that of chimpanzees. The switch from prehistory to history in archaeology is usually associated with the onset of writing and the start of events being recorded. BCE/CE or BC/AD terminology is often used to date 'how old' something is in archaeology rather than 'years ago'. It can be difficult to visualise long periods of time particularly in the (non-linear) geological timeline. One way to do this is to consider the age of the Earth in terms of a 24-hour day. The Earth was formed 4.6 billion years ago and so each one-hour period is equivalent to roughly 192 million years. Dinosaurs went extinct 65 million years ago or around 20 minutes from midnight on the 24-hour clock, i.e. 23:39:40. The start of the Quaternary period is at ~23:59:12, controlled use of fire by *Homo erectus* may date to around 1 million years ago or about 23:59:41 and the Neolithic people were at Stonehenge around 5,000 years ago are <0.1 second to midnight on the 24-hour clock.

a particular sample. For example, Aberdeen granite has a silver-grey appearance with a form of mica called muscovite (containing no iron or magnesium in its structure) whereas granite from Shap in the north of England contains large pink orthoclase feldspar crystals giving an overall pinkish colour to these samples. Working out whether a rock is a granite or a limestone or which of the three families it belongs to involves looking for the clues which were left when the rock was formed. The background to the clues and the type of clue to look for are addressed in the following sections.

3.2.1   THREE KINDS OF ROCK

Rocks are grouped into the three broad categories according to how they originated:

1. Igneous rocks are formed from the cooling of molten magma. These rocks can be further classified as either extrusive (volcanic) when molten magma cools and solidifies on the Earth's surface (it is then referred to as larva) or intrusive when the molten magma solidifies or forces its way into other rock beneath the surface of the Earth. Igneous rocks are often regarded as the primary source of material comprising the Earth's surface.

   The minerals present in a rock and the relative proportions are largely dependent on the chemical composition of the magma. A common extrusive igneous rock is **basalt** whereas **granite** and **gabbro** are common intrusive rocks. One of the main differences between extrusive and intrusive rocks is that the latter tend to contain larger mineral crystals as they solidified more slowly from the magma mixture. For example, coarse grains of pale glassy-looking quartz and white or pinkish blocks of potassium and sodium feldspars make up more than 80% of the mineral content of granite with the small amount of mica appearing as black shiny flakes (Figure 3.3).

2. Sedimentary rocks are formed at the Earth's surface from the accumulation of deposits of pre-existing rocks together with material of organic origin which are then subjected to compaction and cementation processes. These rocks are usually arranged in layered beds, although they can be quite dramatically folded and twisted from their original level orientation, and these strata may contain fossils. Sedimentary rocks are sub-divided into clastic, evaporite and organic according to their mode of origin (Figure 3.4).

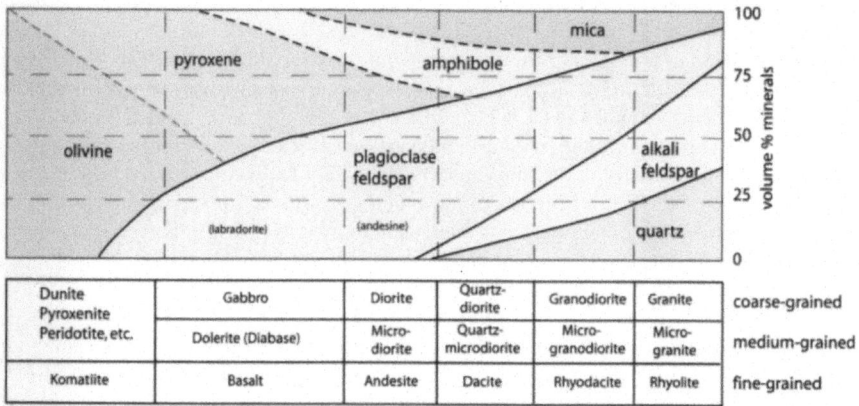

FIGURE 3.3 Classification of igneous rocks based on mineralogy and grain size. (From: Figure 77, page 66. *Rocks and Minerals in Thin Section*, 2nd Edition by W.S. MacKenzie, A.E. Adams and K.H. Brodie. Copyright (© 2017) by CRC Press. Reproduced by Permission of Taylor & Francis Group.)

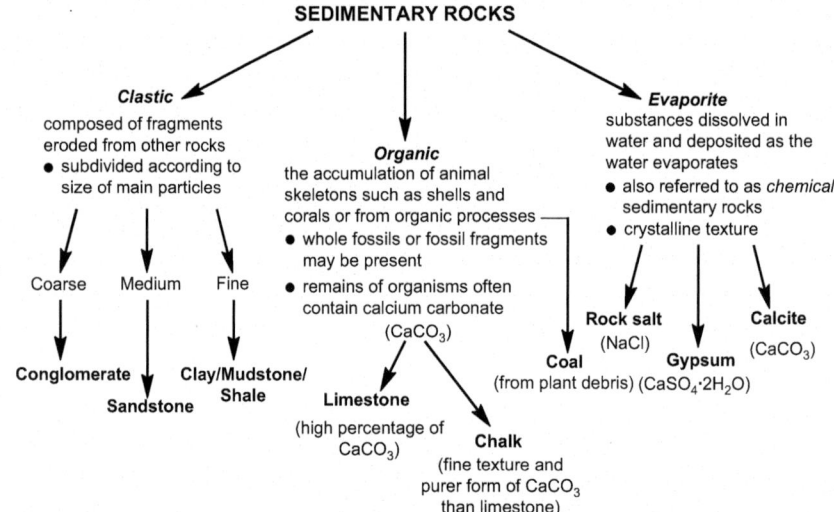

FIGURE 3.4 Some common types of sedimentary rocks.

3. Metamorphic rocks are formed by subjecting any rock type to greater heat and higher pressure from those in which the original rock was formed. The main changes to the rocks are to the mineralogy and structural and textural features whilst the chemical composition remains largely unchanged (Table 3.1). Heating causes recrystallisation of minerals with different shapes, sizes and orientation; pressure can result in foliation (the alignment of minerals in the rock) and fossils and sedimentary structures such as bedding are largely obliterated.

TABLE 3.1

Representative Selection of Common Metamorphic Rocks and the Parent Source Rock

Original Rock		Metamorphic Rock	Notable Features
Sandstone[a]	→	Quartzite	Harder and tougher than original rock, tightly interlocking grains.
Shale[a]	→	Slate	Cleaves into thin sheets in a direction perpendicular to that of the pressure.
Limestone[a]	→	Marble	Recrystallised calcite; white when pure, impurities give a range of colours and patterns.
Bituminous coal[a]	→	Anthracite coal	Hard coal, high carbon content, no fossils.
Granite[b]	→	Gneiss and schist	Minerals in the original rock have been rearranged into bands with alternating light and dark layers. Gneiss has thick layers with coarser mineral grains compared to the finer structure of schist.

[a] Indicates that the parent rock was sedimentary.
[b] Indicates the original rock was igneous.

3.2.2 ROCK AND MINERAL IDENTIFICATION

Initially, the only equipment needed to look closely at rock, mineral or fossil specimens is a small magnifying glass or hand lens. When examining a rock the various bulk characteristics are first considered such as overall colour, texture and size of particles and recording this information in a suitable format. One method of describing a sample is a simple chart such as that given in the Appendix which can be customised according to individual preferences. Alternatively, a notebook or electronic device are equally satisfactory recording media. It can be advantageous to work with known samples and to consult reference material in books, charts or online when first undertaking observations on a rock. This process reveals the similarities and differences between rocks and the variations which can occur within a particular type of rock such as granite or limestone. Making observations on the proportion of different minerals present in a rock and their colour, arrangement, size and shape also helps with identification of the sample. The hardness of a mineral, its resistance to being scratched, can be a diagnostic feature and is found using a simple test. The test involves observing whether objects of known hardness will scratch the mineral and further details can be found in the Appendix.

Some rocks and minerals have particularly diagnostic features, for example, galena (PbS) has a high density and a rock containing this mineral feels heavy when picked up. The presence of calcium carbonate can be detected by adding a few drops of dilute hydrochloric acid or vinegar to a sample of the rock and looking for bubbles (of CO_2). Whether a positive or negative test is obtained is another clue to narrow down the list of possible candidates for the rock or mineral sample. Optical properties of minerals such as transparency and birefringence involve microscopic investigation of thin sections of rock and minerals using specialist laboratory equipment. Information on this aspect of rock and mineral identification can be found in the *Further Reading* section at the end of the chapter and is briefly referred to in the legend of Figure 5.11.

Representative specimens of the most common rocks, minerals and fossils for the type of practical geology outlined above can be purchased inexpensively from rock and mineral shows which regularly take place throughout the country or from online suppliers. With a little experience of observing hand specimens of known rocks, the next step is to take this knowledge into urban or rural environments and to look at these rocks in the field using geological maps, local information on an area or online guides.

3.3 A CLOSER LOOK AT MINERALS

Minerals are naturally occurring inorganic compounds with a definite chemical composition and a crystalline form. Mineral specimens with perfect physical shapes and geometric symmetry can form when conditions are favourable and the crystal can grow unrestricted, but it is more common for crystals to interfere with each other as they grow to form clusters and the aggregate masses seen in rocks.

The vast majority of minerals are compounds of two or more elements and just a few, such as copper, silver and gold, are chemical elements which exist in uncombined forms as a result of their low reactivity. The most abundant elements in the Earth's crust are oxygen (46.1%) and silicon (28.2%) and these two elements combine to form the silicate ion, SiO_4^{4-}. Silicate minerals make up around 90% of the rock-forming minerals with most of the remaining being carbonates $\left(CO_3^{2-}\right)$ and oxides (O^{2-}). Quartz (SiO_2) is one of the purest and one of the most widely distributed minerals and is used in the ceramics and glass making industries and as an ore for elemental silicon. High-quality quartz is required for the manufacture of electronics and this is often obtained synthetically using a hydrothermal method which mimics the natural crystallisation process.

Although the formula SiO_2 of quartz suggests that it is an oxide of silicon it is classified as a silicate as it consists of a three-dimensional network of SiO_4 tetrahedra. In the non-crystalline state SiO_2 forms sand and sandstones but when crystallised into quartz it forms six-sided crystals (Figure 3.5). This shape is an external expression of the internal geometric arrangement of the constituent atoms and thus the symmetrical features of a crystal are useful when identifying minerals. The relationship between the external shape and internal structure of some carbonate and silicate minerals is explored further in the following sections.

3.3.1 Carbonates and Silicates

3.3.1.1 Carbonates

The two most stable, and common, crystalline forms of calcium carbonate are calcite and aragonite. Calcite occurs as rhombohedral crystals with well-defined cleavage planes (the tendency of a mineral to break in preferred directions) whereas aragonite crystals do not have cleavage planes and have pseudo-hexagonal symmetry (Figure 3.6a). Calcite and aragonite adopt quite distinct crystal shapes as a

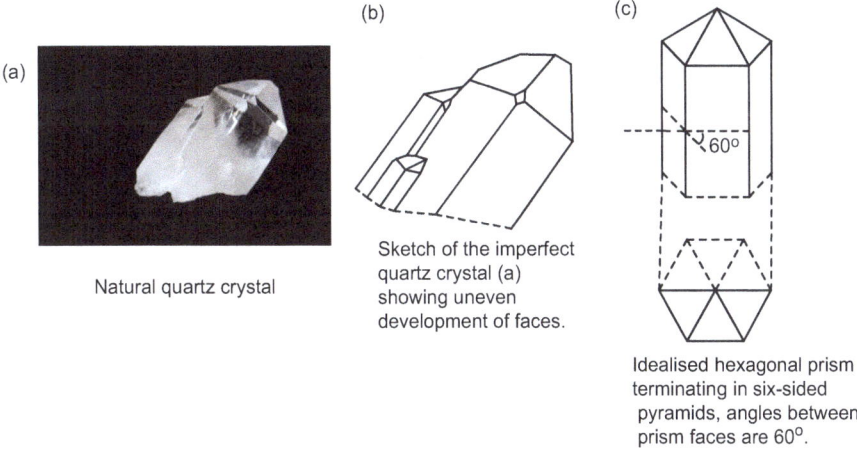

FIGURE 3.5   Quartz crystals: natural and idealised.

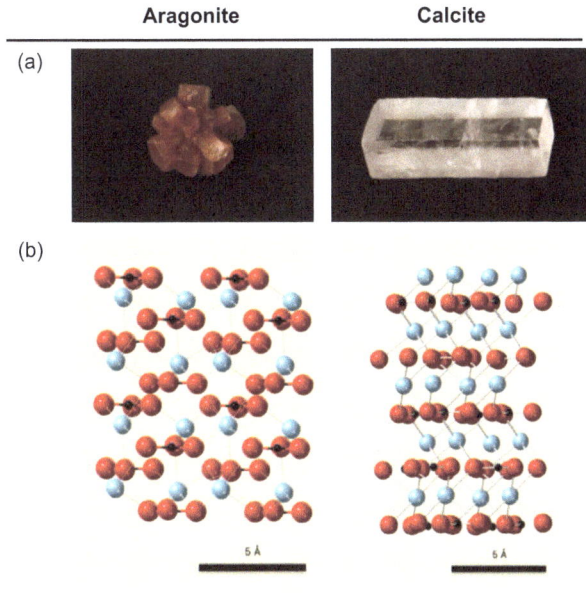

FIGURE 3.6   Aragonite and calcite polymorphs of calcium carbonate. (a) Crystal form and (b) internal arrangement of ions. The blue spheres are calcium ions and the black and red spheres are the carbon and oxygen atoms of the carbonate ions. Crystal structures in (b) were produced using CrystalViewer® 11.2.2. (CrystalMaker Software Limited, Begbroke Science Park, Oxfordshire, UK.)

consequence of the regular (although different) internal arrangement of the calcium and carbonate ions in their lattice structures (Figure 3.6b), and as such they are referred to as polymorphs of calcium carbonate. Both structures are composed of alternating layers of calcium and the (trigonal planar) carbonate ions along the vertical or c-axis. The location of calcium layers is practically identical in both structures and the major difference between the two minerals is the position of the carbonate groups. In aragonite some of the carbonate ions are raised to form two layers and their orientations within the layers are different. The carbonate ions in calcite in contrast all lie in many parallel planes, with each plane running right across the crystal. The interactions between the parallel layers of ions are weaker in calcite than aragonite leading to a weakness in this plane of the crystal. The consequence of this is that calcite breaks easily along these layers whereas aragonite does not. The preferential crystal growth along the c-axis gives rise to the more needle-like morphology in aragonite.

3.3.1.2 Silicates

Silicates are the most abundant group of minerals but have the most complex chemical composition. The fundamental unit is the SiO_4^{4-} anion in which four oxygen atoms form single bonds to a silicon atom at the centre of a tetrahedron (Figure 3.7a). The silicate anion combines with metal cations such as Mg^{2+} and Fe^{2+} to form neutral compounds such as in the olive-green mineral olivine $(Mg,Fe)_2SiO_4$. The comma in the formula between magnesium and iron indicates that these two ions can substitute for each other. Thus, although the mineral has a definite geometrical structure it has a chemical composition ranging from Mg_2SiO_4 to Fe_2SiO_4. The SiO_4^{4-} tetrahedra can join in pairs by sharing one of the corner oxygens to form pyrosilicates with the formula $[Si_2O_7]^{6-}$ (Figure 3.7b and c). The pyrosilicate ion is found in the rare mineral akermanite, $Ca_2MgSi_2O_7$.

Structures consisting of rings, extended chains, sheets or three-dimensional networks can be formed by combining tetrahedra in different configurations. The ratio of silicon to oxygen increases with increasing polymerisation from 1:4 in structures with isolated SiO_4^{4-} tetrahedra to 1:3 in chain silicates and this trend continues into the more complex combinations. Beryl, for example, with the formula $Be_3Al_2[SiO_3]_6$, is a cyclosilicate (Figure 3.7d) with six tetrahedra linked in a ring whilst the dark coloured mineral augite, $Ca(Mg,Fe)Si_2O_6$, is one of the pyroxene group of chain silicates (Figure 3.7e) commonly found in igneous rocks.

Three major silicate minerals are found in granite: quartz (SiO_2), feldspar (e.g. $KAlSi_3O_8$) and mica [e.g. $K(Mg,Fe)_3(AlSi_3O_{10})(OH,F)_2$] and all have structures resulting from complex polymerisation of the silicate tetrahedra. Mica is a sheet silicate with a sandwich structure (Figure 3.8a) in which an octahedral-complexed magnesium layer is packed between two layers of silicate tetrahedra. Some silicon atoms can be replaced by other atoms such as aluminium, and in mica approximately one-quarter of the silicon sites are substituted, giving a tetrahedral sheet formula of $[AlSi_3O_{10}]_n^{5-}$. Aluminium is in an oxidation state of +3 compared to silicon's +4, and so the additional negative charge is balanced by the presence of K^+ cations located in the interlayer spacing. Furthermore, as Fe^{2+} ions can substitute for Mg^{2+} ions and fluoride (F^-) can substitute for hydroxyl (OH^-) in the octahedral layer, the overall formula of mica is $K(Mg,Fe)_3(AlSi_3O_{10})(OH,F)_2$. The tetrahedral-octahedral-tetrahedral sandwiches are weakly bonded to each other by the interlayer cations whereas there is strong bonding inside the sandwiches. The sandwiches can thus be easily separated along

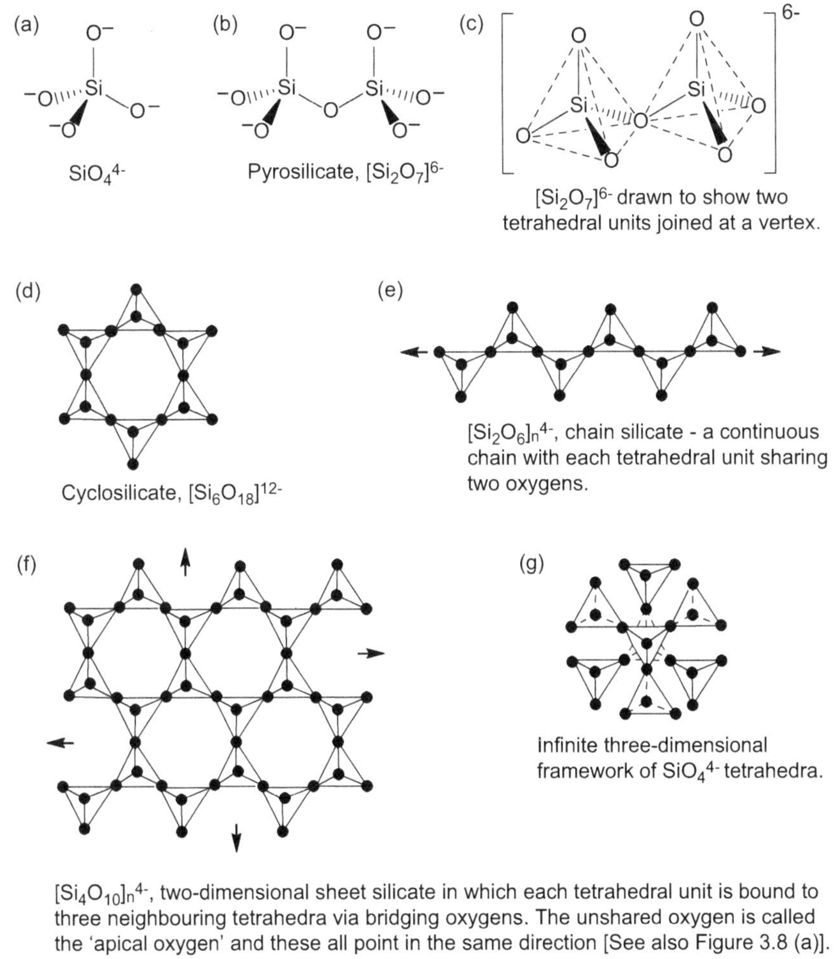

(a)

SiO_4^{4-}

(b)

Pyrosilicate, $[Si_2O_7]^{6-}$

(c)

6-

$[Si_2O_7]^{6-}$ drawn to show two
tetrahedral units joined at a vertex.

(d)

Cyclosilicate, $[Si_6O_{18}]^{12-}$

(e)

$[Si_2O_6]n^{4-}$, chain silicate - a continuous
chain with each tetrahedral unit sharing
two oxygens.

(f)

(g)

Infinite three-dimensional
framework of SiO_4^{4-} tetrahedra.

$[Si_4O_{10}]n^{4-}$, two-dimensional sheet silicate in which each tetrahedral unit is bound to
three neighbouring tetrahedra via bridging oxygens. The unshared oxygen is called
the 'apical oxygen' and these all point in the same direction [See also Figure 3.8 (a)].

FIGURE 3.7 Structure and composition of silicate ions in minerals indicating the way in which silicate tetrahedra are linked together. The oxygen atoms in (d) to (g) are shown as black spheres and silicon atoms are omitted. The negative charge on the silicate anions is balanced by metal cations which can link individual silicate tetrahedra or occupy interstitial sites or cavities within the polymeric structures.

the potassium layers and give mica its characteristic property of easily cleaving into thin sheets.

Both quartz and feldspar are framework silicates (Figure 3.7g) consisting of an infinite network of tetrahedra inter-connected via bridging oxygen atoms. Silicon dioxide, SiO_2, is the only mineral composed exclusively of silicon and oxygen (Figure 3.8b) whereas in feldspar some of the tetrahedra are aluminate, $[AlO_4]^{5-}$, with cations occupying available voids in the structure to balance the charge. In the alkali feldspar series the K^+ and Na^+ can substitute for each other giving compositions ranging from orthoclase ($KAlSi_3O_8$) to albite ($NaAlSi_3O_8$), whereas the plagioclase

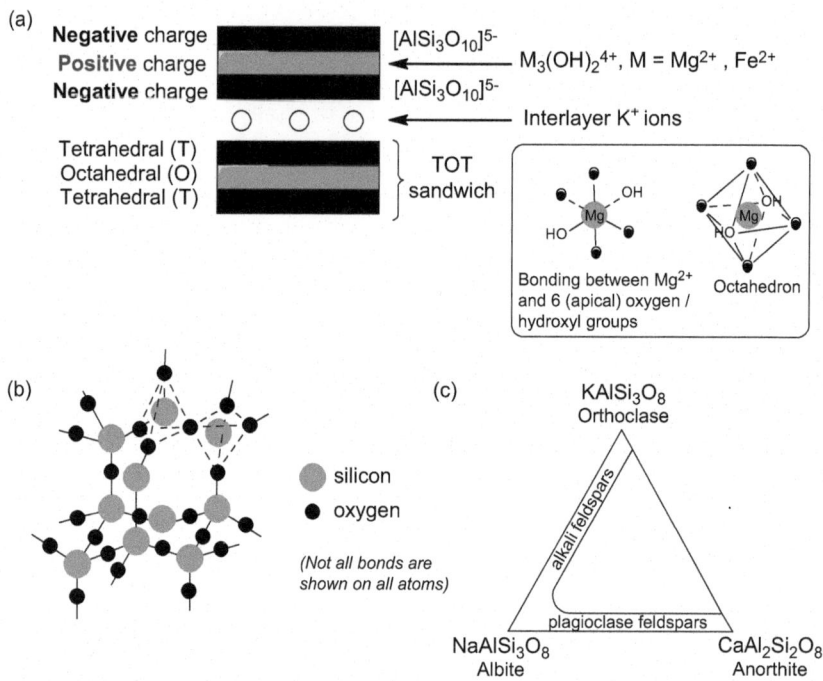

FIGURE 3.8 Major silicate minerals in granite. (a) Tetrahedral-octahedral-tetrahedral (TOT) sandwich structure of biotite mica. Apical oxygens (shown as black dots in the inset diagrams) of the tetrahedral layers and hydroxyl groups coordinate to Mg^{2+} ions forming a layer of octahedra which is the middle of the TOT sandwich. The overall composition for the combined TOT sandwich is $Mg_3Si_4O_{10}(OH)_2$. [Inset: a magnesium ion coordinated to six oxygen atoms or hydroxyl groups gives rise to an eight-sided (octahedral) shape.] (b) SiO_2, part of the structure of quartz showing both silicon and oxygen atoms. Each Si atom is surrounded by four oxygen atoms and each oxygen atom is a bridge connecting the corners of two tetrahedra. (c) Ternary diagram showing the composition of alkali and plagioclase feldspars in terms of the three end members. There is no series of minerals between orthoclase and anorthite because of the large size difference between K^+ and Ca^{2+}. The terms orthoclase and plagioclase are derived from Greek terms for types of cleavage; orthoclase because of the straight (90°) fracture planes (*ortho* = straight and *klasis* = to break) exhibited by the potassium feldspar which produces sharp rectangular edges whereas plagioclase feldspars have cleavage planes which are not exactly at right angles (*plagio* = oblique).

group has the more complex formula of $(Na,Ca)Al_{1-2}Si_{3-2}O_8$. The Ca^{2+} and Na^+ ions have different charges and so a coupled substitution is required to maintain charge balance. The two end members of the plagioclase series are albite ($NaAlSi_3O_8$) and anorthite ($CaAl_2Si_2O_8$). As albite is a member of both series the composition of the feldspars can be represented in a ternary diagram (Figure 3.8c).

The rigid uniform three-dimensional network of Si–O bonds in quartz gives a strong structure and there are no planes of weakness, with the result that this mineral does not show cleavage and when broken the surface is irregular. Feldspars however have some weaker bonds between the metal ions and oxygen and there are two good cleavage planes producing crystals with two flat surfaces which sometimes appear at right angles to each other. The presence of a cleavage helps to distinguish feldspars from quartz. Large pink tabular orthoclase feldspar crystals are a particular feature of Shap granite and this rock features as part of the entrance to London St Pancras station, in St Paul's Cathedral (London) and the plinths of many statues.

Although the main minerals in granite are all complex silicates, it is the detailed structural arrangement of the silicon-oxygen tetrahedral building blocks at the atomic level together with the presence of other elements which accounts for their different physical characteristics and identifying features within a rock environment. Quartz is a colourless and highly transparent mineral but there are many coloured varieties due to the inclusion of small amounts of metal impurities and/ or through a radiation-induced process. Cairngorm or smoky quartz is a brown coloured mineral due to the substitution of Fe^{3+} ions for Si^{4+} in a few tetrahedra with the introduction of monovalent charge compensation cations such as Li^+ in interstitial positions in the lattice. Purple amethyst crystals are the result of Fe^{3+} ion impurities whilst titanium ions or phosphorus substituting as P^{5+} with compensating Al^{3+} ions are believed to give the pink colour of rose quartz. A number of these varieties are valued as semi-precious stones and are sometimes cut in a particular way to take advantage of the mineral's natural planes of cleavage which are determined by its atomic structure.

3.4 METALS AND MINING

Many rocks contain small quantities of metal-bearing minerals but commercially useful amounts usually occur when they are concentrated in veins within other rocks. Many valuable ores were formed when hot watery fluids percolated through rock and dissolved widely dispersed metal minerals as they passed. If the ore-forming solution then encountered an open fracture where rapid flow could occur, the temperature and pressure would drop and this could cause dissolved minerals to be precipitated. Deposition could occur deep underground but the deposit may be uncovered through erosion over millions of years. Important sulphide and oxide minerals occur in this way including chalcopyrite ($CuFeS_2$, a copper ore), galena (lead sulphide, PbS), zinc blende (ZnS) and cassiterite (tin oxide, SnO_2). These ores occur alongside larger quantities of other common (and nonvaluable) minerals such as quartz and calcite which are collectively termed gangue. The recovery of the elemental metal from the ore deposits involves mining operations to bring the mineral to the surface, separating the metal-containing component from the gangue and finally extracting the metal from its compound.

3.4.1 FROM THE COPPER AGE TO THE 20TH CENTURY

The roots of metallurgy stretch back into prehistory when copper implements started to be used by humans instead of those made from stone. Native (naturally occurring) elemental copper was likely to have been used first as it required little or no processing and the discovery that annealing made the metal easier to work probably occurred independently in several early civilisations. At some point early metal workers discovered, possibly by accident, that heating copper ores at a high temperature in the presence of charcoal 'melted out' the copper metal. The liquid metal which fell to the bottom of the hearth, or later a crucible, could be easily collected. Later innovations such as moulding and casting and the discovery of bronze by smelting (the process of heating and melting) a mixture of copper and tin ores enabled people to create metal objects that were hard and durable, anything from jewellery and decorative items to tools and weapons.

The smelting procedure used in modern times involves roasting chalcopyrite in a limited air supply to give Cu_2S and FeO. The latter is removed by combination with silica to form a slag, and Cu_2S is converted into copper (Figure 3.9a). Large quantities of the pollutant SO_2 are produced and since the 1980s alternative hydrometallurgy methods of copper extraction from the ores malachite [$Cu_2CO_3(OH)_2$] and azurite [$Cu_3(CO_3)_2(OH)_2$] have been introduced to avoid the SO_2 emissions (Figure 3.9b). The basic hydrated copper carbonate ores are treated with sulphuric acid to leach out the copper in the form of aqueous $CuSO_4$. The dissolved copper ions are then concentrated, impurities removed through a series of solvent extraction procedures and then the metal is recovered using electrolytic methods. Environmental concerns around mining, rising extraction costs and the move towards renewable energy technologies have led to an increasing proportion of copper used in manufacturing processes being recovered from scrap metal.

The UK has a rich history of mining and the Great Orme copper mine in North Wales dates to around 1,500 BCE. Cassiterite (SnO_2), the principal ore of tin, was found in veins in and around granite areas in Cornwall and Devon and this was the only indigenous source of tin in Britain. The tin and copper ores in the South West were formed closely following a period of mountain building (Variscan orogeny) in the late Palaeozoic era (~300–270 mya) and the mineralisation which produced the North Pennine Orefield is also believed to date from this period. Rocks in this latter orefield are sedimentary,

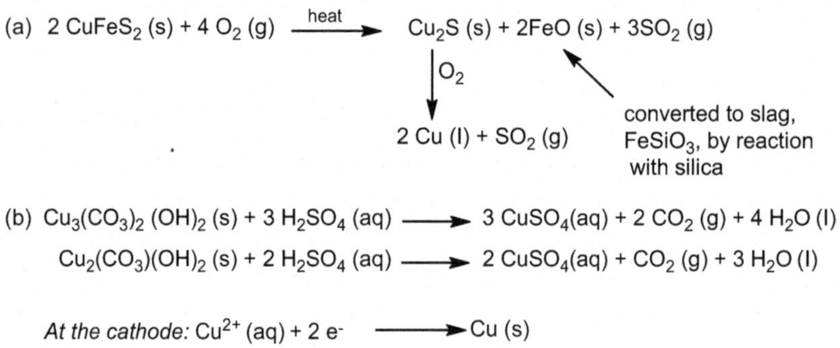

(a) $2\ CuFeS_2\ (s) + 4\ O_2\ (g)\ \xrightarrow{\text{heat}}\ Cu_2S\ (s) + 2FeO\ (s) + 3SO_2\ (g)$

$$\Big\downarrow O_2$$

$$2\ Cu\ (l) + SO_2\ (g)$$

converted to slag, $FeSiO_3$, by reaction with silica

(b) $Cu_3(CO_3)_2\ (OH)_2\ (s) + 3\ H_2SO_4\ (aq)\ \longrightarrow\ 3\ CuSO_4(aq) + 2\ CO_2\ (g) + 4\ H_2O\ (l)$

$Cu_2(CO_3)(OH)_2\ (s) + 2\ H_2SO_4\ (aq)\ \longrightarrow\ 2\ CuSO_4(aq) + CO_2\ (g) + 3\ H_2O\ (l)$

At the cathode: $Cu^{2+}\ (aq) + 2\ e^-\ \longrightarrow Cu\ (s)$

FIGURE 3.9 Extraction of copper metal from ores using pyrometallurgy and hydrometallurgy. (a) A scheme summarising the many reactions that take place during the smelting of chalcopyrite. The metal that is produced in this process is not very pure and electrolysis methods are used to refine the copper. (b) The initial step in the leach-solvent extraction-electrowinning process of malachite and azurite and the final electrolytic step to produce copper metal.

and when hot silica-rich and therefore acidic solutions passed through limestone beds, the calcium carbonate dissolved and the resulting $\left(CO_3^{2-}\right)$ ions then recombined with other ions to produce new carbonate compounds such as siderite ($FeCO_3$). Silica partially replaced limestone (in a process of silicification) over quite large areas and in the process cavities developed into which minerals were later deposited.

Cornish tin was very important for the production of bronze and pewter alloys, not just in the UK but traded across Europe. Mining in Cornwall and the extraction of lead and zinc minerals (galena, PbS and zinc blende, ZnS, respectively) in particular from the North Pennine Orefield had their heyday in the 18th and 19th centuries. Small-scale mining in both these areas is known to have occurred since medieval times, and possibly earlier. Early mines were shallow affairs but as these sites became worked out the miners turned to parent lodes underground and miles of tunnel networks were created in search of valuable mineralised veins. The importance of tin in the medieval period was recognised by special legal documents, the Stannary Charters in 1201 and 1305. These charters granted tin miners in Devon and Cornwall the right to prospect and work tin ore deposits on anyone's land and in return for these privileges they paid a special tax to the crown called 'coinage' on all the tin produced.

Once the ore had been mined it was processed to obtain the metal. The first step was crushing the ore to a fine powder and then washing with water to separate the heavier ore from the lighter gangue material. Physically separating the metal-containing and non-metal components continues to be the initial procedure in most extractive metallurgical methods. The concentrated ore was then roasted in a furnace to convert metal sulphides to metal oxides (and to remove impurities) which were converted to the metal using a carbon-reduction process (Figure 3.10a). The smelting procedures were carried out on an industrial scale in the 18th to 20th century to process the large quantities of ores being mined. The basic chemistry is the same in modern extraction processes although technological and engineering advances have optimised the procedures.

The technology used to smelt copper, tin, lead and zinc ores with charcoal could not produce high enough temperatures to melt iron ores to give liquid metal. Iron could be produced by this method but the product was an impure spongy solid which was worked by the blacksmith to remove the slag (waste) and purify the metal. Medieval iron mining and smelting works were small scale, occurred in many areas of the country and mainly used local iron ore deposits such as limonite [$FeO(OH){\cdot}nH_2O$ in the Forest of Dean, Gloucestershire] and haematite (Fe_2O_3 in Furness and West Cumberland, now part of Cumbria) (Figure 3.10b). Improvements in the chemistry of smelting using coke from coal, engineering innovations in furnace design and the invention of the steam engine all contributed to the rapid expansion of the iron industry in the 18th century. The varied environmental conditions which existed in Britain between the Carboniferous and Jurassic eras (~360–145 mya) deposited what was to become the raw materials for the large-scale production of iron which drove the Industrial Revolution. The most important iron ore for the furnaces was ironstone in which the iron occurred as siderite ($FeCO_3$), but this was mixed with clay and/or coal deposits. Despite the relatively poor quality of ironstone (~30% iron content), compared to the oxides it occurred in large quantities and, with coal and limestone deposits also readily available, it was this iron ore which largely stocked the furnaces.

(a) $2 \text{ MS (s)} + 3 \text{ O}_2 \text{ (g)} \xrightarrow{\text{heat}} 2 \text{ MO (s)} + 2 \text{ SO}_2 \text{ (g)}$

M = Pb, Zn $\downarrow$ C (or CO formed in furnace)

$2 \text{ M (l)} + \text{CO/CO}_2 \text{ (g)}$

$\text{SnO}_2 \text{ (s)} + \text{C (s)} \xrightarrow{\text{heat}} \text{Sn (l)} + \text{CO}_2 \text{ (g)}$

(b) *Calcination:* $2 \text{ FeCO}_3 \text{ (s)} + \frac{1}{2} \text{O}_2 \xrightarrow{\text{heat}} \text{Fe}_2\text{O}_3 \text{ (s)} + 2 \text{ CO}_2 \text{ (g)}$

$\text{Fe}_2\text{O}_3 \cdot 2\text{Fe(OH)}_3 \text{ (s)} \xrightarrow{\text{heat}} 2 \text{ Fe}_2\text{O}_3 \text{ (s)} + 3 \text{ H}_2\text{O (g)}$

Smelting: $\text{Fe}_2\text{O}_3 \text{ (s)} + 3 \text{ C (s)/CO (g)} \xrightarrow{\text{heat}} 2 \text{ Fe (l)} + 3 \text{ CO/CO}_2 \text{ (g)}$

FIGURE 3.10 Equations representing the extraction of metals from their ores using pyrometallurgy. (a) Lead, zinc and tin from sulphide and oxide ores. The reducing agent is a mixture of coke and carbon monoxide which is formed in the smelting process. (b) Haematite, Fe_2O_3, is the main iron ore used for smelting in blast furnaces. Other ores such as iron carbonate (siderite) and hydroxides [e.g. goethite, FeO(OH) and limonite, FeO(OH)·nH$_2$O] are first thermally decomposed ('calcined') to form the oxide Fe_2O_3 which is then reduced with carbon or carbon monoxide in the smelting step. The blackband type of iron ores containing FeCO_3 were particularly valuable as they were laminated generally with sufficient coal to make them self-smelting without the need to add more coal. Addition of limestone in the smelting of most ores aids the removal of impurities such as arsenic, phosphorus and sulphur and increases the efficiency of the process. The molten iron (called pig iron) from the furnace is ~94% iron and contains 2%–4% carbon (which makes it brittle) plus small amounts of other elements dissolved in it. Pig iron is further refined before conversion into more workable grades of iron such as wrought iron (< 0.15% carbon) and into steel. One of these refining steps is an oxidation process which converts the carbon to carbon monoxide, and also removes a number of the secondary impurities. Some elements are added back into molten steel before the metal is cast into ingots and solidified, e.g. stainless steel is 74% Fe, 18% Cr and 8% Ni. Recycling is becoming an increasingly important component of metal manufacture and, e.g. 35% of zinc comes from processing galvanised steel in an electric arc furnace.

It was the fortuitous juxtaposition of raw materials, the innovative metallurgical developments of the Darby family and transport infrastructure which made the mass production of cast iron economically feasible. In 1781 the first bridge to be built of cast iron was opened and this 'Iron Bridge' is a symbol of the Industrial Revolution which originated in the nearby area of Coalbrookdale, Shropshire. The large swathe of Jurassic ironstone deposits of the East Midlands Shelf, an arc running from Bristol via Banbury, Kettering and Scunthorpe to Cleveland in the north-east coupled with a similar dispersion of coal, rapidly gave rise to the iron and steel and metal manufacturing industries in these regions. Scottish iron production principally used carboniferous clayband and blackband ironstones which were included within the Coal Measure deposits of Lanarkshire and Ayrshire.

Coal was a crucial resource not only for generating the coke which acted as the reducing agent within the furnace but also as the fuel to heat the ovens for the pyrolysis procedure to make the coke. Coal was formed in the UK from the Devonian period onwards although most was formed during the Carboniferous period when the land mass lay in warm tropical waters near the equator. Post-carboniferous events and earth movements which have folded and faulted strata resulted in coalfields in South Wales and Scotland often occurring in basin-shaped synclines whilst an eroded arch-shaped anticline has given the series of discontinuous coal seams east of the Pennines. As shallow surface deposits became exhausted, the deeper coals were only able to be extracted through underground mining, often at great depths, with the associated infrastructure to bring it to the surface.

The mining of metal ores declined in the later part of the 19th century and became uneconomic due to increasing cost of extracting material from ever greater depths and the consequent need for more pumping to keep the mines dry and the competition from cheaper imports. Imported ores sustained the metal extraction and manufacturing industries in the 20th century although there was a trend towards sites with coastal locations near sea ports. Coal, which used to generate a significant proportion of Britain's electricity, has been phased out with the switch to a low-carbon economy.

Remnants of former mines such as ruined buildings, smelting chimneys, pit head machinery and railway sidings are slowly becoming absorbed into the landscape. Parts of some older mines have been preserved and can be visited, and it is sometimes possible to find samples of the commoner minerals on old spoil heaps.

3.4.2 21st-Century Opportunities

Gold continues to fascinate. Gold, because of its density, is often associated with streams as it separates from other minerals during weathering and transport and small quantities have been found in all countries of the UK. A find in the Kildonan Burn in Sutherland started a short-lived gold rush in 1869 and current miners are aiming for the economic extraction of gold ore from deep within the hills near Tyndrum in Scotland.

Whilst mining for metallic ores has largely ceased in the UK some mines have continued the extraction of industrial minerals such as barytes ($BaSO_4$) and fluorspar (fluorite, CaF_2). These compounds have a role in heavy industries; fluorspar to remove impurities during iron and steel production and the very dense and soft properties of barytes make it an excellent weighting agent in the drilling mud for oilfields. Barytes and fluorspar are also used indirectly for the manufacture of speciality chemicals and are in a range of household products, from paints to toothpaste. Most of the gypsum ($CaSO_4 \cdot 2H_2O$) mined is for the building industry (dehydrated gypsum is plaster of Paris) and the salt mines in Cheshire have been used for centuries for supplying cooking salt and, more recently, for producing rock salt for gritting roads in winter. A large void in the mine created by the extraction of salt has been recently repurposed for the storage of important documents and paintings.

With the drive towards clean and sustainable energy there is increasing demand for batteries to power electric vehicles and for renewable energy storage systems. Lithium ions (Li^+) from lithium carbonate (Li_2CO_3) or lithium hydroxide (LiOH) are vital for battery manufacture and have been traditionally extracted from underground brine or hard rock ore deposits. Lithium is only a very minor constituent of the multi-component solutions of underground brine deposits, and extraction methods are based on solar evaporative concentration and a series of chemical precipitation reactions. These methods, which are lengthy and laborious, take advantage of the higher aqueous solubility of LiCl compared to NaCl and KCl and that Mg^{2+} ions directly precipitate in the presence of CO_3^{2-} and OH^- ions. These primary separation steps generate a concentrated solution (~3–6 wt%) of lithium salts which is then refined to remove the remaining bulk and minor impurities. Addition of sodium carbonate to the higher-purity concentrate precipitates lithium carbonate which can then be converted into lithium hydroxide by reaction with lime (Figure 3.11a). Extensive research is currently being undertaken to improve the recovery of lithium salts from underground brine deposits, which are largely found in South America, to reduce the environmental footprint associated with this traditional methodology and to develop procedures applicable to less concentrated lithium-brine sources.

Lithium-bearing minerals are largely phosphates and complex aluminosilicates, the most important of which is spodumene, $Li_2O \cdot Al_2O_3 \cdot 4SiO_2$, which occurs in granite pegmatite rocks. Mining of lithium ores, predominantly in Australia and China, is similar to conventional operations, and separation and removal of gangue components gives the spodumene concentrate. The spodumene in the concentrate is in the natural α-form in which Li^+ ions occupy cavities between chains of Si-centred tetrahedra and Al-centred octahedra and cannot easily be extracted. The more porous, and more chemically reactive, β-structure is obtained through a calcination step and the Li^+ ions are then extracted in the form of lithium sulphate using an acid-leaching process. The lithium sulphate can be converted into lithium carbonate via a precipitation reaction with sodium carbonate or reacted with sodium hydroxide to form lithium hydroxide (Figure 3.11b).

(a) $2 Li^+ (aq) + CO_3^{2-} (aq) \longrightarrow Li_2CO_3 (s)$

$\downarrow Ca(OH)_2 (aq)$

$2 LiOH (aq) + CaCO_3 (s)$

(b) *Calcination:* α-spodumene $\xrightarrow{heat}$ β-spodumene

Leaching: $Li_2O \cdot Al_2O_3 \cdot 4SiO_2 + H_2SO_4 \xrightarrow{heat} Li_2SO_4 + Al_2O_3 \cdot 4SiO_2 + H_2O$
β-spodumene   water soluble

Precipitation: $Li_2SO_4 (aq) + Na_2CO_3 (aq) \longrightarrow Li_2CO_3 (s) + Na_2SO_4 (aq)$

FIGURE 3.11   Lithium extraction. (a) Later reactions in lithium extraction from brines. The lithium hydroxide produced in the reaction is initially obtained as the hydrate which is dehydrated by heating under vacuum. (b) Chemical processing reactions in the sulphuric acid method for extraction of lithium from the hard rock mineral spodumene.

The UK currently has no commercial lithium mining industry but several sites in Cornwall are actively under investigation for this critical raw material. Direct lithium extraction (DLE) techniques are being explored and developed to recover lithium from the geothermal briny waters beneath old tin and copper mines near Redruth whilst drilling and extraction of the mineral zinnwaldite, [$KLiFeAl_2Si_3O_{10}(OH,F)_2$], from mica in granites near St Austell requires the development of new processing chemistry procedures. Both mining operations are aiming to deliver lithium carbonate in an environmentally responsible manner.

3.5 FOSSILS

3.5.1 THE NATURE OF FOSSILS

Fossils are the preserved remains of ancient plants and animals and can be found in sedimentary rocks. The oldest known fossils, dating back to around 3.5 billion years, are of cyanobacteria which are detected in the form of stromatolites (layers of bacteria and trapped sediment). Most fossils however date from the Cambrian period, ~550 million years ago, and their study (palaeontology) can give information on the Earth's past environments and climate, provide evidence of evolutionary change and determine the relative age of rocks. Some fossil molluscs, corals and echinoids are very similar to modern species whilst other classes such as ammonites, trilobites and graptolites (Figure 3.12) are now extinct but were very successful for many millions of years.

Ammonites, trilobites and graptolites were small and abundant animals which evolved rapidly such that rock zones containing their fossilised remains can be quite thin. Recognising and interpreting these index fossils contained within the rock strata were pivotal in formulating the relative ages of rocks and their sequence in time in the 19th century and also in correlating rocks which may be widespread geographically. This work led to the construction of a stratigraphic column in which rocks were ordered according to their age with the oldest rocks at the bottom. Assigning accurate dates to different time periods was done in the 20th century using radiometric methodology and these data are shown in Figure 3.13 column (c).

| Ammonite (Jurassic) | Trilobite (Ordovician) | Graptolite (Ordovician) |

FIGURE 3.12 Fossil remains of extinct animals. The term fossil comes from the Latin *fossilis* meaning 'dug up'. Ammonites take their name from medieval Latin *cornu Ammonis* the horn of Ammon, due to the supposed resemblance to the involuted horn of the deity Jupiter Ammon. The 'tri' in trilobite refers to the three longitudinal parts of the body and graptolite is derived from the combination of the Greek *graptos* and *lithos* meaning writing and rock respectively. A 1 cm^3 cube next to the fossils gives an indication of their size.

Boundaries between the different eras and periods in the geological timescale represent significant changes in the history of the Earth and to the life-forms that existed at specific times (see Figure 3.13). The fossils are a record of how these different organisms evolved and adapted to their environment in the past. A widespread and rapid decrease in the biodiversity on Earth signalled a mass extinction, the aftermath of which created new opportunities for surviving species to diversify and evolve as ecosystems recovered. What caused the mass extinctions is the subject of much debate but multiple causes are thought to have contributed to these events. However, changes to atmospheric and oceanic chemistry and climate change, possibly associated with a trigger event such as powerful volcanic eruptions or asteroid impact, have been implicated in the five largest mass extinctions around the ends of the Ordovician, Devonian, Permian, Triassic and Cretaceous periods.

3.5.2 Fossilisation Processes

The process of preserving an organism as a fossil is governed by two main factors: the environment where it died and the materials which made up its body when it was alive. In order to become fossilised the organism must be covered in sediment such as mud or sand soon after death. This protects the remains from scavengers or from being broken up and so species which lived in seas or swamps are more likely to be preserved than those which lived on land. Once the remains are buried their decomposition slows down due to a lack of oxygen. However, it is usually only the hard parts of the remains, which are more resistant to decay compared to the soft tissues, that survive to become preserved as fossils.

3.5.2.1 Permineralisation and Petrification

The hard parts of animals such as the shells, bones and teeth are mostly made of inorganic compounds, for example, calcium carbonate and calcium phosphate, but they are porous and also contain some organic matter and as this decays it leaves cavities. The pore spaces and cavities can become infilled with minerals, usually silica or calcite, which crystallise or precipitate out of the ground water as it percolates through the remains. This process can be taken one step further in the fossilisation of tough woody parts of plants when all the original organic material and the pore spaces can be replaced with minerals. Fossils formed through permineralisation (partial mineralisation) and petrification (complete mineralisation) have the same shape as the original material but wood and bone in particular will be heavier and, depending on the conditions of fossilisation, even the most delicate structures can be preserved (Figure 3.14).

3.5.2.2 Replacement, Moulds and Recrystallisation

Chemical alteration can also occur to the hard parts of an organism's remains. As mineralised ground water solutions move through an exoskeleton, the original inorganic compounds can be dissolved and replaced by another mineral. Replacement by silica, known as silicification, can aid the extraction of fossils from a limestone matrix; the matrix can be dissolved in acid leaving behind the shell-like material. If the organism was buried in an anaerobic sediment rich in iron sulphide (FeS_2, also known as iron pyrites) then this pyritisation process leads to golden coloured fossils as exemplified by the graptolite image in Figure 3.12.

Era or Aeon	Period or Epoch	Age (in millions of years)	Succession of Life	Major Geological Events in UK Landmass
Caenozoic (Age of young life)	Quaternary — Holocene	0	UK becomes permanently occupied	Present day
		0.01		Quaternary Ice Age ends
	Quaternary — Pleistocene	2.6	Evolution of *Homo* genus	Warm climate which cooled as country drifted northwards. Several ice ages. Essentially modern arrangement of continents.
	Tertiary — Pliocene	5	Emergence of hominin species in Africa	
	Tertiary — Miocene	23	Mammals and birds become common	General retreat of sea. Areas of folding and faulting in south as a consequence of Alpine orogeny.
	Tertiary — Oligocene	34	Evolution and diversification of many mammals and birds, flowering plants develop.	Cycles of marine and non-marine conditions in southern England, formation of London clay.
	Tertiary — Eocene	55	Extinction of many groups of animals including ammonites, dinosaurs and marine reptiles at end of Cretaceous, associated with asteroid impact.	Last volcanic rocks formed (Skye, Mull and Giant's Causeway).
	Tertiary — Palaeogene	65	*Age of mammals*	
Mesozoic (Age of middle life)	Cretaceous	145	Reptiles dominant, dinosaurs common, flowering plants emerging. Variety of fresh water, estuarine and marine life.	Initially shallow lagoonal water restricted to south and east England, formation of Purbeck limestone. Salt water advances later in period with first greensand and then chalk deposition.
	Jurassic	200	Flora and fauna abundant, including terrestrial dinosaurs, first birds and new marine creatures evolve, ammonites common. Mass extinction cleared the way for dinosaurs to assume dominance on land. *Age of reptiles (including dinosaurs)*	Warm and shallow sea advances, and then retreats, across central and eastern England, many island areas. Widespread depositions of clay, mudstones, limestones and sandstones.
	Triassic	250	Reptiles common, first dinosaurs evolved. Half of all marine species gone, including trilobites and many corals and brachiopods decimated.	Dry semi-arid environment during Permo-Triassic period, desert sandstones formed, marine fauna limited and evaporite deposits in northern England from drying up of inland salt lakes.

FIGURE 3.13 Chronostratigraphic Timeline [Columns (a) to (c)] and Summary of Important Events in the Geological [Column (f)] and Biological [Columns (d) and (e)] History of the UK.

(*Continued*)

(a)	(b)	(c)	(d)	(e)	(f)
Palaeozoic (Age of old life)	Permian	300	Age of amphibians	First reptiles on land, amphibians important in wetter areas. Conifers become important.	Further tectonic collisions caused Variscan orogeny ~290 mya affecting SW England, with granite intrusions and mineralisation events creating Cornish and Pennine orefields.
	Carboniferous	360		Abundance of life in sea and on land including corals, brachiopods and ammonites. Forests of giant ferns and horsetails. Insects abundant and large. Extinction event marked by a plummeting in the evolution of new species.	Widespread sea shelf over large part of British area which is near equator, climate tropical with wet deltaic swampy forests. Coal measures deposited, further volcanic activity in central Scotland.
	Devonian	415	Age of fishes	Amphibians developed from fish. Rapid colonisation of the land. Seed forming plants appear. Diverse sea life; fish widespread, many reef-building corals but triobites declining, graptolites became extinct. Earliest insects in fossil record.	Sea in the south, mountains undergoing rapid erosion with debris forming deltas and swamps draining into sea. Terrestrial deposits known as Continental Old Red Sandstones formed.
	Silurian	445		Corals, and brachiopods flourished. Terrestrial life established and vascular plants evolved. Mass extinction caused a dramatic decline of marine life.	Geosyncline silted up, Iapetus Ocean finally closes, 'soft docking' of Scotland and England producing mountain ranges in Southern Uplands and further south. Granite intrusions in Highlands. Shallow marine shelf in southern and eastern areas.
	Ordovician	490	Age of invertebrates	Primitive fish-like creatures develop. No animal life on land. Primitive plants move onto land. Diverse marine invertebrates.	Formation of a geosyncline or basin as the 'America' and 'Europe' plates became closer with shallow seas in other areas. Onset of Caledonian orogenesis in north and west Britain.
	Cambrian	550		Development of marine fauna with hard shells; triobites became dominant, varied brachiopods and gastropods start to appear. Graptolites (floating colonial animals) common.	England and Wales lay near the South Pole and climate cold. Scotland was joined to North America. A shallow sea covered much of the area.
Proterozoic	Precambrian (> 85% of the Earth's history)	2,500		First plants ~700 mya Land fungi appeared ~1.3 billion years ago. Multicellular life developed ~1.5 billion years ago.	Supercontinent situated around south Pole starts to break up into individual plates ~600 mya. Formation of Torridonian sandstones ~1200–800 mya ago.
Archaen		4,000		Single-celled life only.	Metamorphosis of ancient rocks into Lewisian gneisses ~2.6 billion years ago. Earth's crust and atmosphere developing.
Hadean		4,600			Formation of the Earth.

FIGURE 3.13 (Continued) Chronostratigraphic Timeline [Columns (a) to (c)] and Summary of Important Events in the Geological [Column (f)] and Biological [Columns (d) and (e)] History of the UK.

FIGURE 3.14   Cut and polished fossils; (a) solitary coral and (b) ammonite.

When a calcium carbonate shell is buried and then subsequently dissolved out of its enclosing hardened sediment, it leaves an impression or external mould in the rock of its outer surface. Alternatively, if the shell becomes infilled with minerals, particularly silicates from surrounding sediments or ground water, and then the calcium carbonate material is dissolved away, an internal mould of the shell's inner surface is left on the outer surface of the secondary material. When external moulds become infilled with minerals then a cast of the original organism is formed. Casts can also be created synthetically using plaster of Paris or latex rubber to generate a replica of an original fossil. This procedure is often used to make replicas of rare fossilised skeletons of dinosaurs, for example.

Many molluscs, including the extinct ammonites, use the aragonite polymorph of calcium carbonate to make their shells rather than calcite. Aragonite is the less stable polymorph and these shells, under the pressure of burial in sediment, can undergo a recrystallisation process to the more stable calcite crystalline form.

3.5.2.3   Carbonisation and Compression

Plant leaves and stems which were rich in cellulose are often preserved as a thin film of carbon. As the original buried plant material becomes overlain with more and more sediment the weight of these layers causes the oxygen, hydrogen and nitrogen of the organic material to be gradually distilled off leaving behind a film of carbon. As the original organism is squashed flat these carbonised fossils are also known as compression fossils. Graptolites (see Figure 3.12), an extinct group of colony-forming marine animals, are often found in shales as flat compression fossils that look like pencil markings with a 'tuning-fork' or saw blade appearance. During graptolite fossilisation the external layer of chitinous material sometimes became mineralised with chlorites, a group of sheet silicates, or partly pyritised resulting in both pale and black fossils.

3.5.2.4   Fossilisation without Alteration

Insects first started appearing in the Devonian period and became common around 200 mya but they can be delicate with few hardened structures and most are terrestrial. As a result they have a poor fossil record and those fossilised in the usual methods are mostly incomplete. However, small whole insects and spiders have been entrapped in amber (solidified tree resin) when it evolved in the Eocene, around 50 mya. The insects became trapped in sticky sap from coniferous trees and then this resin hardened in a polymerisation process over time encasing the insect in amber.

Hard parts of remains sometimes resist weathering and chemical action and can be preserved with little or no physical change, such as calcitic shells and teeth from younger Mesozoic and Tertiary rocks. Unaltered organic soft parts are rarely preserved but frozen remains of animals such as woolly mammoths from the Quaternary period have been recovered from the Siberian permafrost.

3.5.3   LOOKING AT FOSSILS

A hand lens, reference information and fossil samples are, as with rock and mineral identification (Section 3.2.2), the key resources to examine and make observations. Common fossils can be purchased inexpensively and rare species can be seen in local natural history museums and a chart such as that given in the Appendix is useful to record findings. Local palaeontology clubs are also very valuable sources of information and advice.

3.6   GEOCHRONOLOGY

3.6.1   BIOSTRATIGRAPHY AND RELATIVE DATING

Biostratigraphy uses fossils to establish the relative ages of the rock strata with numerical ages being established through one of more of the absolute dating methods, many of which are based on the measurement of the decay of radioactive isotopes (see Figure 3.15 for an overview of the various methods). Fossils are used as time markers in relative dating and each individual stratum of sedimentary rock can be identified by the group of fossils it contains. Furthermore, the presence of a distinctive group of fossils also enables the correlation of equivalent beds in different locations.

The relative ages of the sedimentary strata are determined by the principle of superposition: younger rocks rest upon older ones. However, folding, faulting and other processes which operate on rocks that have already formed can disturb the original order and complicate the determination of a sequence of relative ages. Fundamental principles are also applied to igneous rocks such as the law of intrusive junctions: if an igneous rock has been intruded into a sedimentary stratum, or into another igneous rock, then the intruded rock must be the younger. An example of this type of process is included in the interpretation of the rock sequence (**Site 2**) in Figure 3.16.

The seven rock types in Figure 3.16 are numbered according to their relative ages with the oldest layer (unit 1) containing the remains of ancient trilobites and graptolites whilst the youngest (unit 7) has a collection of recent fossils. Five rock units (1, 2, 3, 5 and 7) are sedimentary with their relative ages determined by the law of superposition whilst the principle of intrusion enabled the placement of the two igneous rocks (4 and 6) into the sequence of geological events. The volcanic ash in **Site**

Uranium series dating
Family of methods using different unstable uranium isotopes that decay into stable lead isotopes by different chemical pathways.

^{234}U–^{230}Th (an unstable daughter isotope) used to analyse precipitated $CaCO_3$, e.g. stalagmites which have precipitated from water containing trace amounts of uranium from ~1,000 years ago to 350,000 (500,000 upper limit).

^{235}U–^{207}Pb and ^{238}U–^{206}Pb is used to analyse zircon in rocks formed from ~1 million to 4.6+ billion in regions that are not volcanically active.

^{40}K–^{40}Ar dating: volcanic rocks from 100,000 to 4.6+ billion years ago.
Radioactive decay of unstable isotope of potassium, ^{40}K, into ^{40}Ar (a gas) which becomes trapped within mineral crystals. Half-life of this process is 1.25×10^9 years and the ratio of ^{40}K : ^{40}Ar in the mineral sample gives the time since the mineral began to trap ^{40}Ar. Used to date volcanic rocks containing potassium minerals such as feldspars and micas.

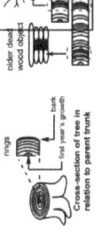
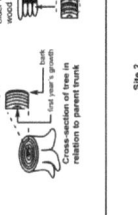

$^{40}_{19}K$ electron capture $^{40}_{18}Ar$ neutron neutrino proton

Radiocarbon (C–14) dating: organic materials only from 500 to 70,000 years ago
^{14}C isotope is unstable and when an organism (plant or animal) dies it stops taking in CO_2 to use to build its tissues and the concentration of ^{14}C in its body starts to decrease through radioactive decay. Knowing the half-life of ^{14}C, the age of the organism can be calculated by measuring amount of ^{14}C left in the sample.

$^{14}_{6}C$ β decay $^{14}_{7}N$ + e proton neutron

Dendrochronology: exact date of timber materials from present day to ~10,000 years ago
Trees produce a ring of new wood annually. The variation in cross-sections of wood can produce exact timelines. Tree-ring chronologies are built by overlapping ring patterns from successively older samples. Useful for samples that are too recent for radiocarbon dating. The tree-ring timescale is also used to calibrate several millennia of the radiocarbon timescale. Tree-ring widths are influenced by climate and this data can also be used to study present climate and to reconstruct climates of the past millennium.

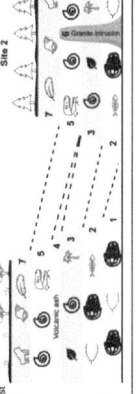

annual rings — last year's growth — both Cross-section of tree in relation to parent trunk

older dead wood object dead tree living tree bark Chronology building

Biostratigraphy and index fossils: from 10,000 years ago to Cambrian
Younger sedimentary rocks are on top, older rocks on bottom, fossils and artifacts found in those layers can be determined as older or younger in time. Index fossils are specific plants or animals that are characteristic of a specific span of geological time and can be used to date sediments in which they are found. Sediments that are deposited far apart but contain the same index fossil species then represent the same limited time period.

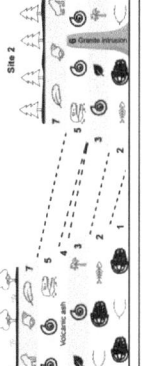

Site 1 Site 2 Youngest rocks Oldest rocks volcanic ash Granite intrusion

ABSOLUTE

RELATIVE

Millions of years ago: 4,600 6 5 4 3 2 1

Years ago: 500,000 400,000 300,000 200,000 100,000 50,000 10,000 1,000 0

FIGURE 3.15 Relative and absolute dating methods. Dating rocks based on which fossils are present is a relative method, all the other methods shown give numerical data and thus are referred to as absolute methods. The boxed areas show the time period for which the various methods give data.

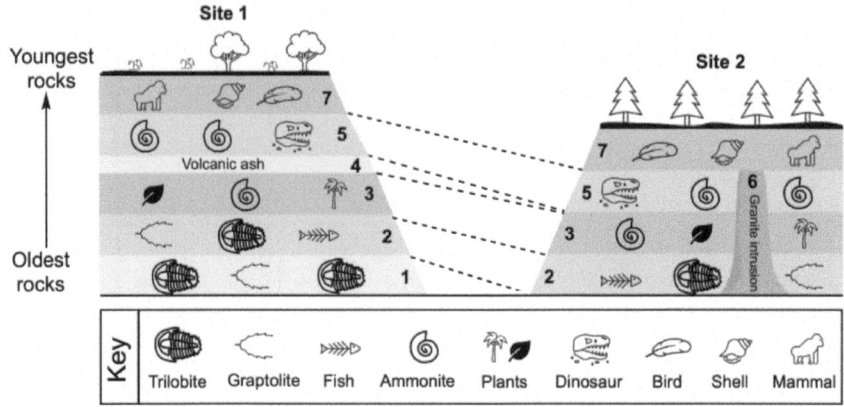

FIGURE 3.16 Idealised biostratigraphy and hypothetical illustration of the use of fossils as biological markers to sequence rock units and to correlate these units in different localities.

1 was deposited onto sedimentary layer number 3 and is thus younger than this rock unit whilst the granite in **Site 2** is younger than the strata into which it intrudes but older than the overlying layer containing distinctive bird and mammal fossils.

3.6.2 ABSOLUTE DATING USING RADIOISOTOPE METHODS

Fossils however cannot give the absolute age of rocks expressed in years; this is done by the analysis of radioactive decay of atoms in certain minerals. One or more isotopes of some elements are unstable and decay by emitting radioactive particles from the nucleus of each of its atoms and as a result become transformed into atoms of another element. The rate at which a specific radioactive parent element decays into a daughter element is constant and can be expressed in terms of the half-life of the parent: the time required to reduce the number of parent atoms to one-half (Figure 3.17).

Quantifying geological time requires isotopes with long half-lives which are nevertheless short enough to allow the collection of data on both parent and daughter elements. The radioactive isotope of potassium, ^{40}K, and the two radioactive uranium isotopes, ^{238}U and ^{235}U, with half-lives of 1.25, 4.50 and 0.710 billion years respectively, meet these criteria and have been used to establish an absolute timescale for the palaeontological record. If, for example, the granite intrusion and volcanic ash of the idealised biostratigraphy diagram in Figure 3.16 were shown to have 94% and 75% of the parent ^{235}U isotope remaining, then by using the decay graph these rocks would be approximately 70 and 300 million years old respectively. The half-life of ^{14}C is geologically short (5,730 years) and is useful for dating materials up to ~50,000 years old and this technique is discussed in greater detail in Section 3.7.2 in the context of archaeological specimens.

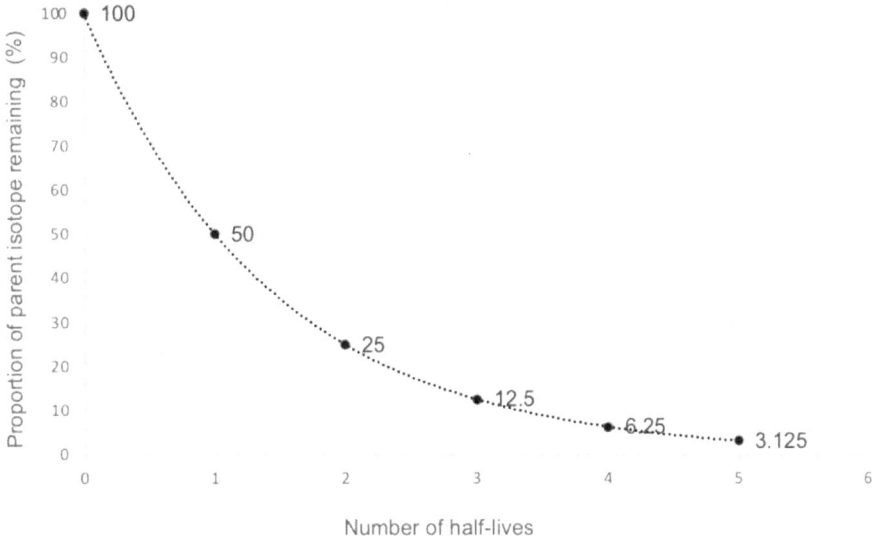

FIGURE 3.17 Exponential decay of radioactive parent element into a daughter element. For example, if there were 64 radioactive atoms to start with then after one half-life there would be 32 atoms left, after 2 half-lives 16 atoms left and so on. The decay graph can be used to calculate the approximate age of a sample with a more accurate figure being determined using the radioactive decay formula which is given in Figure 3.22.

3.6.2.1 Potassium-Argon

The potassium-argon (K/Ar) system is particularly useful for the dating of igneous rocks such as alkali feldspar ($KAlSi_3O_8$) and biotite mica [$K(Mg,Fe)_3(AlSi_3O_{10})$ $(OH,F)_2$]. Decay of the radioactive isotope of potassium, ^{40}K (0.012% or one atom out of every 8,600 atoms), is a dual process forming two daughter products: ^{40}Ar through an electron-capture process and ^{40}Ca through β^- emission. Although this latter is the more common branch (89.5%), the small amount of ^{40}Ca produced through the radiogenic process is difficult to measure as ^{40}Ca is the abundant, stable isotope of calcium. Only the 10.5% of ^{40}K which decays to ^{40}Ar therefore has a role in age determination (Figure 3.18).

Argon is a gas and it escapes from molten larva but becomes trapped when the rock solidifies and over time the argon accumulates in the crystal lattice. The age of the rock sample is calculated from the equation given in Figure 3.19. The K/Ar technique is often combined with a related $^{40}Ar/^{39}Ar$ method which enhances the accuracy of the date obtained. Metamorphic rocks may also be radiometrically dated but, as some of the argon may escape during metamorphism, the technique typically yields the date of the metamorphic event rather than the age of the rock itself. Radioactive minerals often do not occur in sufficient quantities in sedimentary rocks and sedimentary strata are dated by being bracketed between two bodies of igneous rocks that have been dated radiometrically. For example, the age of stratum 5 in the biostratigraphy example in Figure 3.16 would be between the dates obtained for

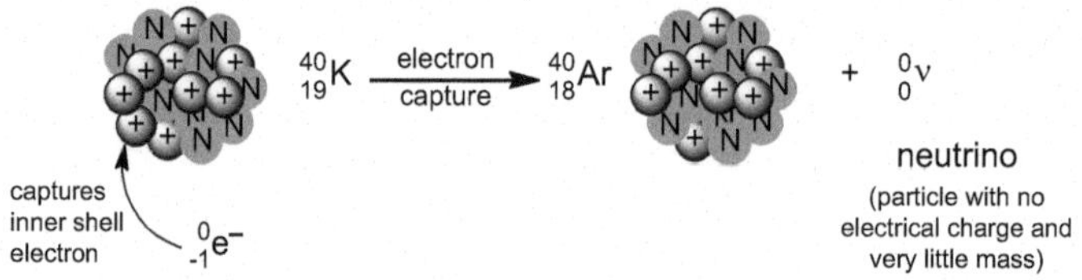

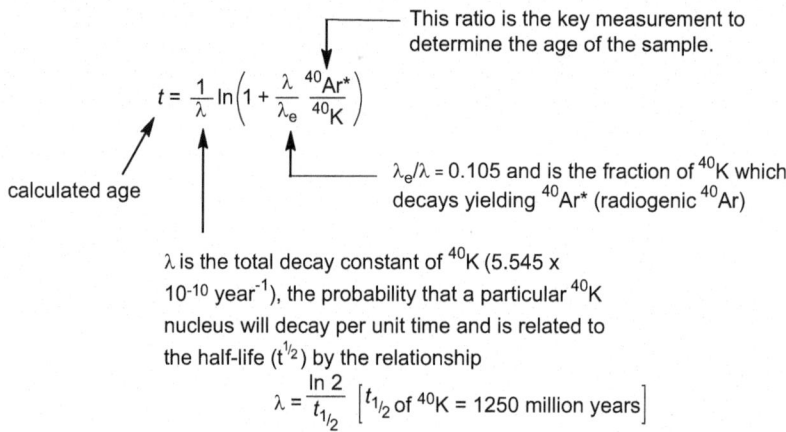

FIGURE 3.18 Radioactive decay by electron capture. An electron is captured by a proton in the nucleus creating a neutron with the ejection of a neutrino. The mass number (number of protons plus neutrons) is unchanged and the atomic number (number of protons) is reduced by one, thus generating a different element.

$$t = \frac{1}{\lambda} \ln\left(1 + \frac{\lambda}{\lambda_e} \frac{^{40}Ar^*}{^{40}K}\right)$$

This ratio is the key measurement to determine the age of the sample.

calculated age

$\lambda_e/\lambda = 0.105$ and is the fraction of ^{40}K which decays yielding $^{40}Ar^*$ (radiogenic ^{40}Ar)

λ is the total decay constant of ^{40}K (5.545 × 10^{-10} year^{-1}), the probability that a particular ^{40}K nucleus will decay per unit time and is related to the half-life ($t^{1/2}$) by the relationship

$$\lambda = \frac{\ln 2}{t_{1/2}} \quad \left[t_{1/2} \text{ of } ^{40}K = 1250 \text{ million years}\right]$$

FIGURE 3.19 Age equation to determine how old a rock is from potassium-argon data. Data for obtaining the amounts of radiogenic argon-40, $^{40}Ar^*$, and ^{40}K is determined separately. Crushing and heating mineral samples (under vacuum) releases the trapped argon, and the amount of ^{40}Ar present is determined by mass spectrometry. The proportion of non-radiogenic ^{40}Ar in the sample is calculated using the known $^{40}Ar/^{36}Ar$ ratio of atmospheric argon and then this is subtracted from the total ^{40}Ar to obtain the radiogenic $^{40}Ar^*$ amount. The total potassium content of the sample is measured by atomic absorption spectroscopy and, as the present day $^{40}K/K$ ratio is constant, the amount of ^{40}K in the sample can be determined.

the intrusive igneous rock and the extrusive layer of volcanic ash. Furthermore, the increased sensitivity of the analytical instrumentation in recent years has resulted in these techniques finding a greater role in the archaeological field.

3.6.2.2   Uranium-Lead Methods

Two isotopes of uranium are used, ^{238}U and ^{235}U, and these decay into ^{206}Pb and ^{207}Pb respectively, and the nuclear reactions which lead to the stable lead isotopes result from the expulsion of many α and β particles (Figure 3.20). Dates obtained from the two different decay series can be correlated which acts as a quality check on the reliability of the result.

Zircon ($ZrSiO_4$) is the key mineral used in uranium-lead dating as it occurs in a wide range of rock types. Trace amounts of uranium and thorium ions substitute

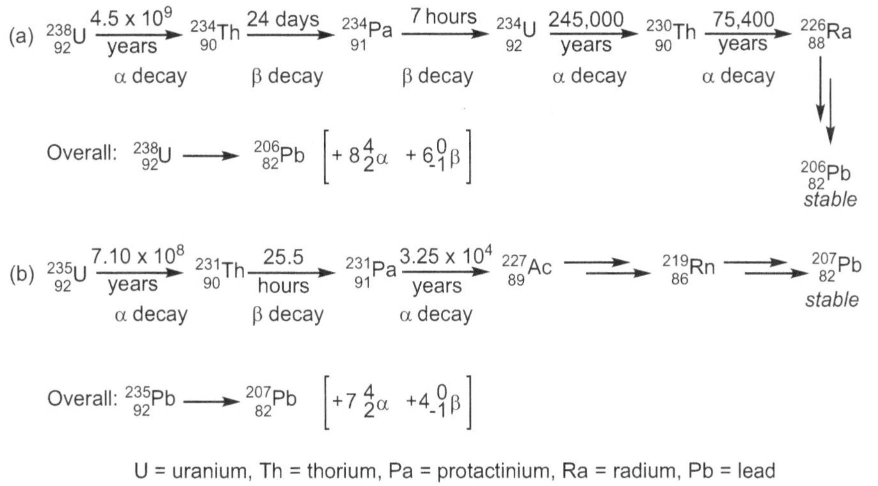

(a) $^{238}_{92}U \xrightarrow[\alpha\ \text{decay}]{4.5 \times 10^9 \text{ years}} {}^{234}_{90}Th \xrightarrow[\beta\ \text{decay}]{24 \text{ days}} {}^{234}_{91}Pa \xrightarrow[\beta\ \text{decay}]{7 \text{ hours}} {}^{234}_{92}U \xrightarrow[\alpha\ \text{decay}]{245,000 \text{ years}} {}^{230}_{90}Th \xrightarrow[\alpha\ \text{decay}]{75,400 \text{ years}} {}^{226}_{88}Ra$

$$\downarrow$$

Overall: $^{238}_{92}U \longrightarrow {}^{206}_{82}Pb \left[+ 8{}^{4}_{2}\alpha + 6{}^{0}_{-1}\beta \right]$

$^{206}_{82}Pb$ stable

(b) $^{235}_{92}U \xrightarrow[\alpha\ \text{decay}]{7.10 \times 10^8 \text{ years}} {}^{231}_{90}Th \xrightarrow[\beta\ \text{decay}]{25.5 \text{ hours}} {}^{231}_{91}Pa \xrightarrow[\alpha\ \text{decay}]{3.25 \times 10^4 \text{ years}} {}^{227}_{89}Ac \longrightarrow {}^{219}_{86}Rn \longrightarrow {}^{207}_{82}Pb$ stable

Overall: $^{235}_{92}Pb \longrightarrow {}^{207}_{82}Pb \left[+7{}^{4}_{2}\alpha +4{}^{0}_{-1}\beta \right]$

U = uranium, Th = thorium, Pa = protactinium, Ra = radium, Pb = lead

FIGURE 3.20 Uranium decay sequences starting from (a) ^{238}U and (b) ^{235}U isotopes. Each α-decay lowers the atomic number by two units and the mass number by four. β-emission does not change the mass number but raises the atomic number by one. α-Radiation is composed of helium nuclei (He^{2+}) and β-radiation is composed of electrons (e^-), both of which possess high kinetic energy. The energy associated with the radiation is transferred to any material used to stop the particles or absorb the radiation. Nuclear equations generally do not show the charge of the radiation emitted.

for Zr^{4+} in the structure at the time of crystallisation from magma, but Pb^{2+} ions are too large to be incorporated into the lattice and thus the amount of non-radiogenic lead in zircon crystals is vanishingly small. Zircon is present in very small amounts (an accessory mineral) in many igneous rocks and, being a resilient mineral, it persists in sedimentary deposits and is a common constituent of most beach and river sands. Furthermore, the presence of zircon in detrital deposits is partly due to its high density, a property which also aids the separation of this mineral from other lighter materials present in sand or rock samples. Electron microscope images of zircon crystals can show growth rings which represent different periods of crystallisation and reflect the geological history of the mineral sample. The high melting point and physical robustness of zircon enable this mineral to survive several tectonic and metamorphic events such that selective microsampling of domains within crystals can give insight into these thermal episodes occurring over a period of thousands or millions of years. Mass spectrometry methods are again central to obtaining numerical data on the amounts of uranium and lead isotopes present in a sample.

3.7 ARCHAEOLOGICAL CHRONOLOGY

The Quaternary period of geological time covers the last 2.6 million years up to the present and during this period the Earth's climate has fluctuated between large-scale glaciations that persisted for up to 100,000 years and shorter warm intervals. The Quaternary period is also largely synonymous with the period of archaeological time (see Figure 3.2) associated with the evolution of the *Homo* genus. Although *Homo sapiens* (modern humans) is the only surviving living species of the *Homo* genus, there is fossil evidence for more than 20 species of ancient human relatives. Stone tool making and use has been associated with the emergence of the genus *Homo*, the oldest representative of which is *Homo habilis* (2.4 to 1.4 million years ago). This species is believed to be descended from the genus *Australopithecus* and human evolution itself extends back to around 6 million years ago in the Tertiary period of the Caenozoic era, when our lineage split from that of the chimpanzees.

Archaeological evidence suggests that there have been several periods of human occupation in Britain over the last million years but continuous permanent settlement did not occur until the climate improved at the end of the last Ice Age around 12,000 years ago. Most archaeological sites contain organic material and so the most common method for determining their age is radiocarbon dating. Stone artefacts of the early humans are not dated directly but are linked to the age of the sediment layers in which they are found. Both the carbon-14 and sediment dating methods are considered in this section together with strontium isotope analysis which is used in prehistory migration and mobility studies.

3.7.1 ^{234}U-^{230}TH DATING AND THE PONTNEWYDD CAVE SITE

Pontnewydd Cave near St Asaph in Denbighshire (North Wales) was first excavated in the 19th century and produced flints, animal remains and a large hominin tooth (now lost). Seventeen hominin teeth were found deep inside the cave during modern excavations between 1978 and 1995 and one, an adult molar, resembles those of known Neanderthal fossils and an age of around 200,000 years was determined from a combination of uranium-thorium and thermoluminescence dating.[2]

Neanderthals (*Homo neanderthalensis*) were one of our closest extinct human relatives; they evolved in Europe and Asia whilst *Homo sapiens* evolved in Africa. The Neanderthals lived from about 400,000 to 40,000 years ago, were stocky in appearance and were adapted to the cold. Their teeth show a feature of taurodontism, where the roots of the molars are merged for much of their length unlike the molars of modern humans. Taurodontism may have conferred an advantage to people whose teeth were subjected to substantial wear on the chewing surface. The life of an ordinary molar will end when the crown is worn down but in a taurodont tooth the roots can fill with secondary dentine, allowing it to function for longer.

Archaeology survived at Pontnewydd Cave because sediments at the entrance became laden with water thus slumping them deeper inside together with bones and debris left by human activity. In their new position these deposits were protected from ice advances in the valley below. This mudflow was further safeguarded by layers of stalagmites, forming over it during periods of particularly warm and wet climate and sealing the sediments beneath. A stalagmite formation just above the layer in which the tooth was found was dated to 180 ± 20 kyr and so the Neanderthal remains must be at least this old. Thermoluminescence data on a burnt flint from the same layer of rock deposit as the tooth gave a date of 200 ± 25 kyr and this artefact is also evidence for the use of fire at the site.

3.7.1.1   Principles of ^{234}U-^{230}Th Dating

The decay of uranium-234 to thorium-230 is part of a much longer decay series beginning in ^{238}U and ending in ^{206}Pb (see Figure 3.20). Due to the short half-lives of ^{234}Th and ^{234}Pa in comparison to the timescales being considered, both isotopes can be ignored in ^{234}U-^{230}Th dating and the decay sequence can be reduced to

$$^{238}_{92}\text{U} \xrightarrow[\text{years}]{4.5\times10^9} \, ^{234}_{92}\text{U} \xrightarrow[\text{years}]{245,000} \, ^{230}_{90}\text{Th} \xrightarrow[\text{years}]{75,400} \, ^{226}_{88}\text{Ra} \qquad (3.1)$$

The ^{234}U-^{230}Th dating technique takes advantage of the difference in water solubility between uranium and thorium compounds; uranium compounds are highly soluble whilst thorium compounds are extremely insoluble. As water percolates through the bedrock, soluble uranium compounds will dissolve in the passing water and when calcite precipitates to form stalagmites it incorporates some trace uranium and will have a Th:U ratio close to zero. With time, ^{230}Th starts to appear as the ^{234}U isotope decays. The initial build up is at a uniform rate but as ^{230}Th is itself radioactive, the rate of accumulation of this nuclide decreases and eventually the production rate of ^{230}Th equals its decay rate and a constant value is reached. This is known as secular equilibrium and the ratio of the decay rate (activity) of the daughter nuclide (^{230}Th) to the parent (^{234}U) nuclide is unity (Figure 3.21). The state of secular equilibrium occurs because the half-life of the daughter radionuclide is much shorter than the half-life of the parent. It usually takes several half-lives of the daughter nuclide for the equilibrium to be established and this also dictates the upper age limit which can be obtained from this technique. Mass spectrometric measurements of the amounts of ^{230}Th and ^{234}U isotopes from the sample and calculation of the activities of these nuclides thus allow the time since the calcite stalagmite precipitated to be determined.

Stalagmites and other cave mineral deposits (collectively known as speleothems) are palaeoclimate archives; their growth rings retain information on the climatic conditions occurring when the different layers of calcite were deposited. Coral fossils are also natural recorders of climate and their growth histories can give information on oceanic conditions in the past. Samples taken along the growth axis of the stalagmite or fossil coral are dated using the U/Th technique and the oxygen isotope record is used as a proxy for the climate, i.e. a surrogate for a variable that cannot be measured directly.[3] Palaeoclimate data has hitherto largely been used to construct a narrative for past climatic conditions but as the methodology and instrumentation associated with this field improves, there is scope for indirect proxy information to be incorporated into the development of future computer models simulating the Earth's climate.

3.7.2   CARBON-14 DATING AND MODERN HUMAN ARTEFACTS

Carbon-14 (also known as radiocarbon) dating is the archaeological workhorse for inferring the chronometric age for specimens of biological origin up to around 50,000 years ago. The last 50,000 years (i.e. the late Pleistocene and Holocene periods) broadly correspond to the time when modern humans spread across Europe and replaced the Neanderthal population. Modern humans (*Homo sapiens*) first evolved

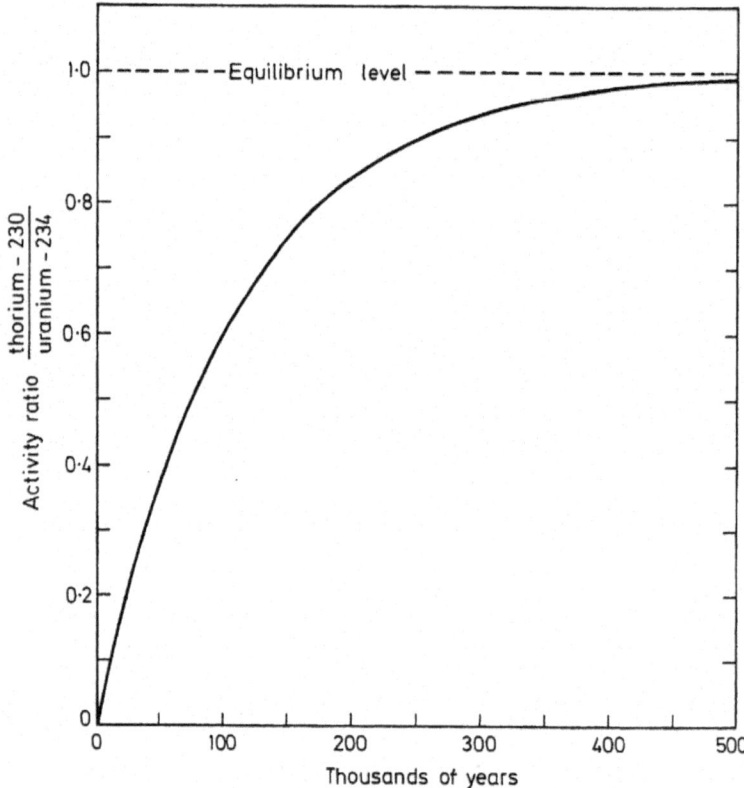

FIGURE 3.21   Basic concept of uranium-thorium dating. The activity ratio on the y-axis compares the decay rate of the ^{230}Th and ^{234}U radioactive nuclides where the activity of each type of nuclide equals $\lambda_x N_x$ where λ_x is the decay constant for nuclide x and N_x is the number of atoms of nuclide x. The curve represents solutions to the equation $[^{230}\text{Th}/^{234}\text{U}]_A = 1 - \exp(-\lambda_{230} t)$ where $\lambda_{230} = (0.693/75.4)$, t is the age in kiloyears, $\exp(x) = e^x$ and $e = 2.718$. Note that this curve assumes the activities of ^{234}U and ^{238}U are the same with the ratio equal to 1. However, the activity ratio is usually greater than 1 which complicates the U-Th age equation and the graphical representation thereof. Details of this more complex age equation and calculations involving the three nuclides, ^{238}U, ^{234}U, and ^{230}Th, together with further information on the technique can be found in sources listed in *Further Reading* at the end of the chapter. (From: Figure 5.3, page 125. *Science-based Dating in Archaeology* by M.J. Aitken, Copyright (© 1990) by Pearson. Reproduced by Permission of Taylor & Francis Group.)

in Africa around 300,000 years ago and dispersed out of the continent in several stages with the current human population outside of Africa thought to be descended from migrants who left approximately 70,000 years ago.[4]

Continuous permanent occupation of Britain started to appear after the end of the last Ice Age and Star Carr in North Yorkshire is regarded as one of the most important and informative Mesolithic hunter-gatherer sites. Many antler, bone and wood artefacts were preserved in waterlogged peat and radiocarbon dating and archaeological evidence suggests that the site, which extends for nearly two hectares, was occupied on a seasonal basis intermittently over several centuries starting around 9,300 BCE. There is evidence of built wooden structures and the timber platforms have given insight into early methods of carpentry using stone tools. Particularly intriguing are the hollowed-out deer skull headdresses, for which the site is most famous, and a decorated pendant is the oldest known piece of art.

The Orkney archipelago is world-renowned for its Neolithic landscapes and architecture. The best-known sites are the Ring of Brodgar, Stones of Stenness, Maeshowe and Skara Brae which are all close together on the largest island. The latter two structures were built from sandstone blocks quarried and shaped from the local bedrock and this type of well-laid and dressed stone construction is also being uncovered at the recent excavations at the Ness of Brodgar. Excavations at this three-hectare site on a headland between the Stones of Stenness and the Ring of Brodgar stone circle reveal a complex of imposing buildings containing a range of artefacts from carved stones and tools to decorative Neolithic artwork. Radiocarbon dating of organic material found at the site suggests that its heyday was from around the late 4th to the middle of the 3rd millennium BCE, although there appears to be earlier constructions beneath those currently excavated. These dates are broadly in line with those recorded for Ring of Brodgar, Stones of Stenness, Maeshowe and Skara Brae and suggest that this area of the archipelago was of pivotal importance to the Neolithic Orcadians.

3.7.2.1 Basis of Radiocarbon Dating

Carbon-14 atoms are constantly being created in the Earth's atmosphere by the interaction of cosmic-ray neutrons with atmospheric nitrogen. Once formed, the carbon-14 atoms combine with atmospheric oxygen to form radioactive carbon dioxide.

$$^{14}_{7}N + \text{neutron} \rightarrow {}^{14}_{6}C + \text{proton}$$
$$^{14}_{6}C + O_2 \rightarrow {}^{14}_{6}CO_2$$

(3.2)

The ^{14}C atoms undergo β-decay but they are being replaced by new ^{14}C atoms at a constant rate and so the ratio of ^{12}C to ^{14}C in the air, and in all living things, at any given time is nearly constant (1 in a trillion carbon atoms are ^{14}C).

$$^{14}_{6}C \xrightarrow{\beta \text{ decay}} {}^{14}_{7}N + e^-$$

(3.3)

Radioactive CO_2 mixes with ordinary CO_2 and is incorporated into plants by photosynthesis and thence by ingestion into animals. During its life the plant or animal has the same proportion of carbon-14 as the atmosphere, as the metabolic processes of the organism ensure that there is constant rapid exchange of carbon between the reservoirs of the carbon cycle. Once the organism dies it ceases to acquire carbon-14, the carbon-14 within its organic remains will start to decay and so the ratio of carbon-14 to carbon-12 will gradually decrease. As carbon-14 decays at a known rate, the proportion of radiocarbon can be used to determine how long it has been since a given sample stopped exchanging carbon with its surroundings; the older the sample the less carbon-14 will be left.

3.7.2.2   How the Age of an Archaeological Sample Is Obtained

The carbon-14 concentration in the sample is measured as a ratio with the carbon-12 (or carbon-13) content using accelerated mass spectrometry. Comparisons with measurements from modern standards of known age are used in order to calculate the amount of radioactive decay that has occurred in the sample and thus obtain the radiocarbon age of the sample using the equation in Figure 3.22. Materials for analysis, which include bones, wood, charcoal, peat, grains, cloth, shells and speleothems, are first pre-treated to remove contaminants. The sample is then burned to convert the carbon content into CO_2 (or graphite) which is then admitted into the mass spectrometer for analysis.

Laboratory measurements on a sample yield a radiocarbon date based on the assumption that the atmospheric ratio of carbon-14 has been constant. However, the amount of ^{14}C in the atmosphere is not stable and has varied over time due to changes in the Earth's magnetic field and solar activity which means that the radiocarbon dates require calibration. Furthermore, the age of a sample is conventionally reported as a calendar date along with a margin of error reflecting uncertainties associated with the calibration and in the experimental data used to calculate the radiocarbon age. To obtain the calendar date the radiocarbon date is compared with internationally agreed calibration curves based on other measures of time such as dendrochronology, uranium/thorium dating of cave stalagmites and corals and samples whose dates are known independently through historical records. Adjustments are also made for man-made atmospheric effects such as the burning of fossil fuels and nuclear testing which decreases and increases respectively the levels of radiocarbon in the atmosphere. The calibration curve itself cannot be described by a formula; there are wiggles and humps and the steepness of the curve varies as illustrated in Figure 3.23.

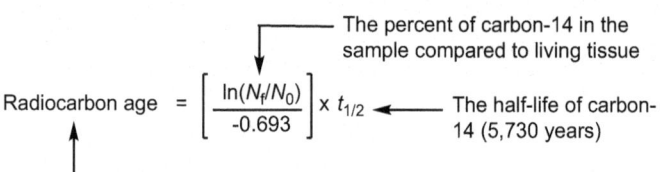

Radiocarbon dates are reported as years before present (BP) to avoid needing to know the date of the analysis. 'Present' is defined by convention as AD 1950, the year when radiocarbon dates were first published.

FIGURE 3.22   Equation to calculate the radiocarbon date of an archaeological sample. For example, if a hypothetical archaeological artefact was found to contain 15% carbon-14 compared to a living sample then its radiocarbon age is 15,700 years before present (BP). This radiocarbon formula is derived from similar equations used in the kinetics of first-order reactions: $\ln(N_f/N_0) = -kt$ (the integrated rate equation), and $t_{1/2} = 0.693/k$ where 0.693 is the value of $\ln 2$ and k is the rate constant. The actual calculation of a radiocarbon date is somewhat more complicated than the equation would suggest as a correction term is required to account for isotopic fractionation because plants discriminate against the heavier isotopes of carbon during photosynthesis, taking up proportionally less carbon-13 and carbon-14 compared to carbon-12.

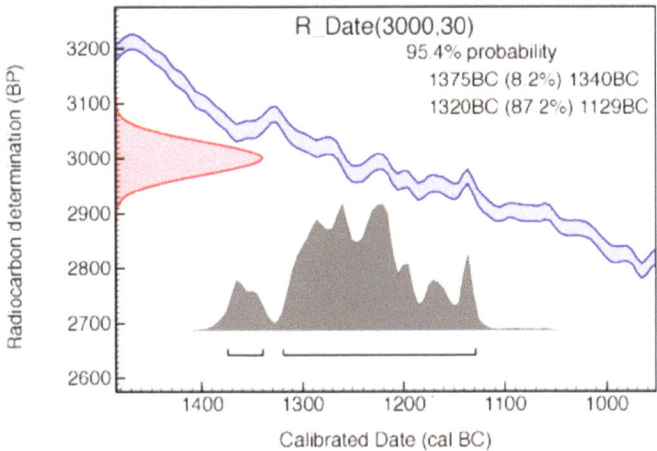

FIGURE 3.23 Calibration of a radiocarbon measurement result of 3000 ± 30 BP. The vertical axis shows the radiocarbon date expressed in years before present (BP, i.e. before 1950) and the horizontal axis shows the calibrated date using BC/AD or BCE/CE terminology. The pair of wiggly lines is the calibration curve (plus and minus one standard deviation) and the symmetrical bell-shaped curve on the left indicate the measured radiocarbon age (plus associated error) in the sample. The intersections between the bell-shaped radiocarbon age and the calibration curve give a probability distribution of possible dates for the sample in the form of a histogram. A range of possible dates is given along with levels of confidence. In this example, there is a 95.4% probability that the sample dates to the period between 1375 cal BC and 1129 cal BC with a 87.2% likelihood that the date falls between 1320 cal BC and 1129 cal BC. (From: (1) University of Oxford (2024) *Oxford Radiocarbon Accelerator Unit.* [Online] Available from: https://c14.arch.ox.ac.uk/. [Accessed: 15th August 2022]. (2) Reimer, P., Austin, W., Bard, E., Bayliss, A., Blackwell, P., Bronk Ramsey, C., Butzin, M., Cheng, H., Edwards, R., Friedrich, M., Grootes, P., Guilderson, T., Hajdas, I., Heaton, T., Hogg, A., Hughen, K., Kromer, B., Manning, S., Muscheler, R., Palmer, J., Pearson, C., van der Plicht, J., Reimer, R., Richards, D., Scott, E., Southon, J., Turney, C., Wacker, L., Adolphi, F., Büntgen, U., Capano, M., Fahrni, S., Fogtmann-Schulz, A., Friedrich, R., Köhler, P., Kudsk, S., Miyake, F., Olsen, J., Reinig, F., Sakamoto, M., Sookdeo, A. and Talamo, S. (2020). The IntCal20 Northern Hemisphere radiocarbon age calibration curve (0–55 cal kBP). Radiocarbon, 62.)

3.7.3   THE BOSCOMBE BOWMEN AND ^{87}SR/^{86}SR RATIOS

Strontium isotope analysis is used as a tool to investigate the geographical origins and movement of people rather than a method of working out how old something is.

3.7.3.1   Basis of ^{87}Sr/^{86}Sr Ratios

Strontium has four stable isotopes (^{84}Sr, ^{86}Sr, ^{87}Sr and ^{88}Sr) and one of these, ^{87}Sr, is radiogenic being produced by β-decay of ^{87}Rb (half-life 48.8 billion years). The amount of ^{87}Sr is expressed as a ratio against the ^{86}Sr isotope and Figure 3.24 shows a map of the ^{87}Sr/^{86}Sr values for Great Britain.[5] The data was obtained by measuring the strontium isotope composition of biosphere components, predominantly plants. The range of values, from ~0.70 to ~0.72, is a consequence of the weathering of

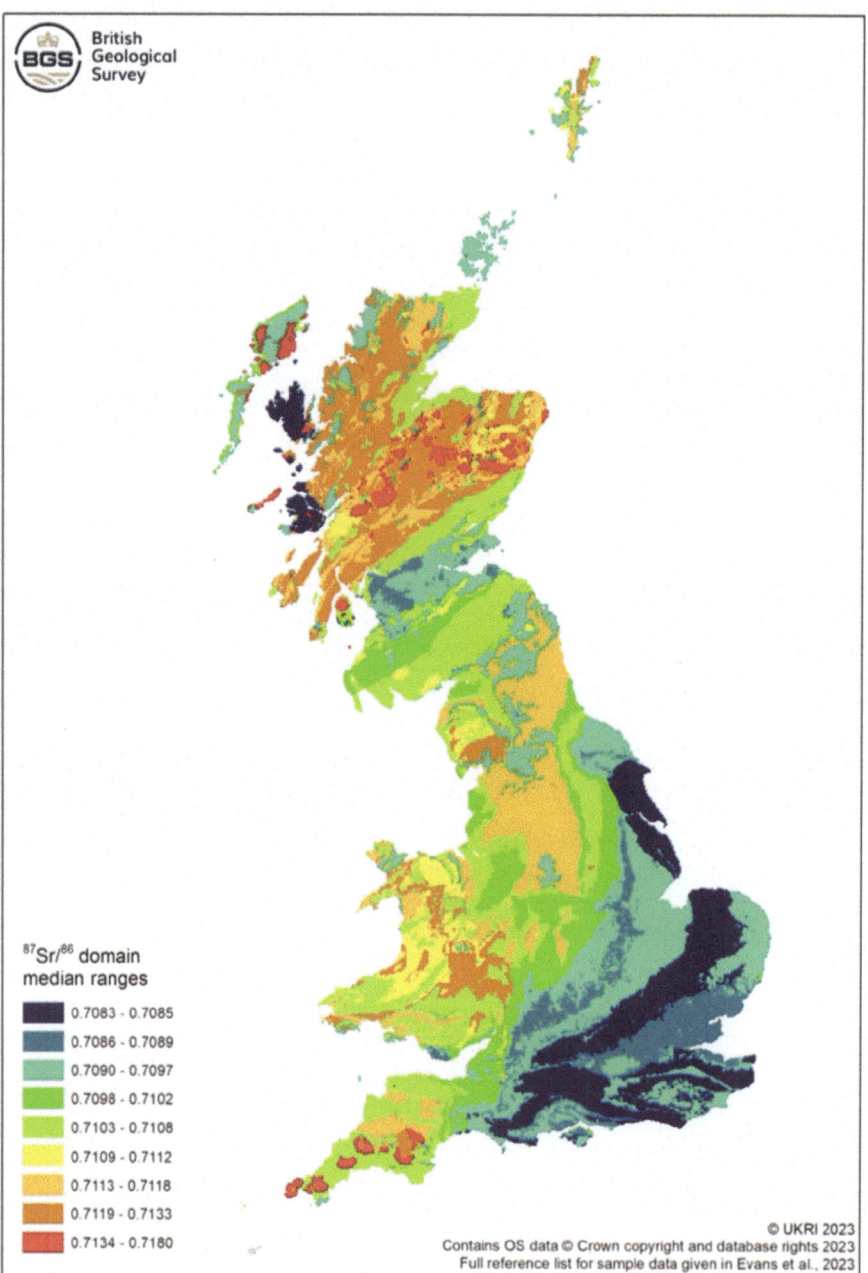

FIGURE 3.24 Strontium isotope ($^{87}Sr/^{86}Sr$) map contains British Geological Survey materials © UKRI [2024] from Biosphere isotope domains (Great Britain). (https://www.bgs.ac.uk/datasets/biosphere-isotope-domains-gb/.)

the different underlying rocks to form soil and strontium enters into the biosphere via plants. The older and more Rb-enriched the bedrock, the more enriched it is in ^{87}Sr. Typically, young/low-rubidium rocks give ^{87}Sr/^{86}Sr values around 0.706 in the biosphere whilst values in the region 0.718 indicate older and/or more rubidium-rich rocks.

Strontium is chemically similar to calcium and soluble strontium compounds are taken up by plants and enter the bones and teeth of humans. Tooth enamel forms during the early years of life. Its composition is not altered subsequently and thus information on the movement of ancient humans across isotopically different terrains during their lifetimes can be obtained from ^{87}Sr/^{86}Sr ratios. Premolars mineralise and therefore lock in the isotope evidence between 3 and 6 years of age, i.e. during early childhood, whereas the third molars mineralise in early adolescence between 9 and 13 years of age. The isotope values for strontium in tooth enamel will therefore reflect the childhood diet and provide clues as to where an individual grew up.

3.7.3.2 Bronze Age Childhood Migration

The Boscombe Bowmen is the name given to a collective burial near Stonehenge from the Early Bronze Age about 2,300 BCE. The grave contained the remains of what is thought to be seven individuals: three adult males, a teenage male and three children, and was discovered in May 2003 during the digging of a trench for a new water pipe. The grave goods included pottery, an antler pendant, a boar's tusk and flints including five barbed arrowheads, the latter giving the name 'The Boscombe Bowmen' to the buried individuals.

Enamel from two teeth from each of the adults (a second premolar and a third molar) and a (unerupted) premolar tooth from each of the two children were analysed together with the dentine component from the third molar from one of the adults. The dentine of the tooth lies directly beneath the enamel and, as this material is more porous, its ^{87}Sr/^{86}Sr ratio reflects the rock from the burial environment. Results from the teeth of the three adults suggest that they spent their early childhood in an area with a radiogenic ^{87}Sr/^{86}Sr isotope signature of ~0.713–0.7135. They each then moved, during early adolescence (as represented by data on the third molar), to a less radiogenic area of rocks, where they acquired an ^{87}Sr/^{86}Sr signature of around 0.7116. As the individuals were buried near Stonehenge on the chalk downland (with a ^{87}Sr/^{86}Sr biosphere value of around 0.708) it implies that they must have travelled at a later time in their lives, to what is now known as Wiltshire. The enamel from the two children yielded ^{87}Sr/^{86}Sr ratios of 0.7097 and 0.7098, which is a very close match, and suggests that they were both raised in the same environment. It is unlikely that the children were raised on the chalk near Stonehenge as their tooth enamel ^{87}Sr/^{86}Sr is significantly higher than the dentine value of 0.708 obtained from one of the adults which reflects the burial environment.

The relatively high ^{87}Sr/^{86}Sr ratios of the adult's childhood can only be matched to a few places in Britain: Cornwall, the Lake District and parts of the Scottish Highlands and Wales. Archaeologists have speculated that Wales is a particularly attractive option as it has known links with Stonehenge at this time.

3.7.4 STONEHENGE

Single standing stones, dolmens, stone circles and chambered cairns are iconic monuments of the Neolithic period and what is visible in the landscape has survived from thousands of years ago. Constructions from wood have rarely survived in the archaeological record unless exceptional circumstances have prevailed to enable their preservation, such as burial in peat which occurred with the Mesolithic site at Star Carr. Some of the aforementioned stone structures are part of a wider network of monuments scattered around the local landscape and constructed over a period of time. Two of the most archaeologically rich UK landscapes are those in Orkney and the Avebury/Stonehenge area in Wiltshire. Both sites feature monuments to the dead (Maeshowe and Stonehenge) and these engineering masterpieces may also have associations with astronomy; the passage in the Maeshowe chambered tomb is aligned with the midwinter sunset and in Stonehenge the sun rises above the Heel Stone on the summer solstice.

Stonehenge is a unique stone circle on account of the architectural design, the large size and shaping of the sarsens and the use of horizontal lintels to lock the upright stones with mortice and tenon joints. The form visible today is the ruin (Figure 3.25) of the monument built around 2,620–2,480 cal BCE although the site itself was originally a cremation cemetery set within a circular ditch and bank earthwork constructed about 500 years earlier.[6] At the centre of the stone monument is a horseshoe arrangement of five sarsen trilithons surrounded by a double arc or arc plus circle of

FIGURE 3.25   Stonehenge as it is seen today.

bluestones with an outer circle of lintelled sarsens.[7] The use of two different types of stone, the bluestones and sarsens, to build this structure on the chalk of Salisbury Plain is a further unique feature of Stonehenge and the provenance of these stones is being pinpointed through geochemical analysis and petrographic examinations.

The sarsen megaliths (hard silicified sandstones) form the primary architecture of Stonehenge and typical uprights weigh on average 20 mt and are generally 4–5 m tall.[8] Elemental analysis of the sarsens was obtained through portable X-ray fluorescence (pXRF), a non-destructive technique, and the results indicated a composition greater than 99% silica for all but two of the stones with traces of the elements Al, Ca, Fe, K, Mg, Mn, P and Ti. This consistent chemistry suggested a common source area for most of the stones. This common source was identified through analysis of data from mass spectrometry and atomic emission spectrometry (destructive techniques) on a core sample from Stone 58 drilled during restoration work in 1958 and a representative range of sarsen boulders from southern Britain. The core sample analysis gave a chemical signature which was representative of the 50 stones with a similar composition from the pXRF data and the best match with the trace element data from 20 different localities was West Woods. This site is 15 miles north of Stonehenge and close to the Marlborough Downs, the long-believed source of the sarsen megaliths. Marlborough Downs is three miles from the Avebury stone circles and is thought to be the source of the naturally shaped megaliths contained within the boundaries of the huge (circumference of 1.3 km) bank and ditch earthwork.

The smaller bluestones, weighing on average two tons, are predominantly igneous spotted dolerites and rhyolites which largely appear to have originated in the Preseli Hills in southwest Wales around 150 miles away from the Stonehenge site. Some have been sourced to specific outcrops within the Preseli area such as Carn Geodog and Craig Rhos-y-felin. Archaeological excavations in the latter outcrop have identified a Neolithic hearth and evidence of activity beside the recess for a rhyolite pillar and where there is a close petrographic match in the adjacent rockface to a Stonehenge bluestone. Radiocarbon dates of 3,500–3,120 and 3,620–3,360 cal BCE of two carbonised hazelnut shells from the Neolithic occupation layer have raised the possibility of quarrying dates in the second half of the 4th millennium BCE and a more complex narrative for some of the bluestones than previously thought.

One of the bluestones is also anomalous, being a sandstone rather than an igneous rock, and historically this stone was believed to have come from Wales along with the other bluestones. The anomalous stone, often referred to as the Altar Stone, lies in front of what would have been the largest trilithon standing at the head of the horseshoe structure. Recent geochemical analysis of the Altar Stone using pXRF indicated that it has an unusually high barium content which effectively ruled out a Welsh origin. Geological studies on the chemical composition and ages of detrital mineral grains such as zircon in a sedimentary rock can effectively give it a fingerprint and it is this methodology which is revealing the original geographical location of Altar Stone. Matching up the fingerprint of Altar Stone with the signatures of sedimentary rocks throughout Britain and Ireland indicates it is very likely to have come from the Old Red Sandstone in the Orcadian Basin in the northeast of Scotland. These rocks are largely of Devonian age and were formed during the closing stages of the Caledonian orogeny.

Whilst scientific analysis of the stones of Stonehenge provides robust evidence for their geographic origins, other aspects such as how the stones were quarried, shaped, transported and erected at the site continue to be more open to interpretation. For example, both overland and sea routes have been proposed, and argued over, for moving the bluestones from Wales to Stonehenge whilst the longer distances and challenging terrain between northeast of Scotland and Wilshire might point more towards a marine transport route. The amount of time and labour invested in the construction of Stonehenge suggests a well-organised society and this type of culture is echoed in the megalithic landscapes in Orkney and around Newgrange in eastern Ireland. Shared architectural elements and rock art motifs between Neolithic monuments in Orkney and Ireland also suggest people were travelling long distances and sharing ideas. Strontium isotope analysis of animal bones at Durrington Walls, the site of a large Neolithic settlement close to Stonehenge, indicates that people were bringing and eating animals from many parts of Britain and that the major feasting event occurred around midwinter. Celebrating midwinter when the sun sets over the Altar Stone at Stonehenge appears to have been a significant point in the year for Neolithic people as it is also marked at Maeshowe in Orkney and Newgrange in Ireland. Passages in the Orkney and Newgrange tombs are aligned with the midwinter setting sun and the midwinter sunrise respectively. One proposal is that Stonehenge became a unifying monument to link widely dispersed communities from many parts of Britain over many generations and that it fell into disuse with the arrival of metal-working people from continental Europe and the beginning of a new type of culture.

NOTES

1 Quartz is a glassy and clear or white mineral with generally irregular crystals. Feldspars are commonly white or pink crystals, often with flat faces. The name is derived from the German for *feld* (field), *spar* (crystal) and reflects the abundance of these minerals. The term 'alkali auxiliary' means the mineral contains the alkali metal ions sodium or potassium but very little calcium in contrast to plagioclase feldspar (see Section 3.3.1.2). Mica appears as very small and black crystals which sparkle in the granite, from Latin for shining.

2 Thermoluminescence dating measures the amount of light released when an object is heated. Crystalline minerals such as feldspar and quartz absorb and store energy from natural radiation emitted from radioactive impurities in the sample. The nuclear radiation causes the excitation of electrons which become trapped in a metastable energy state and this energy is released in the form of light when the sample is heated. When clay is fired or other stones are burnt the accumulated thermoluminescence up to that time is released; in effect the 'clock' is reset to zero. Over time the crystals continue to absorb and store energy until the sample is heated during the measurement process hundreds or thousands of years later and thus the time which has elapsed since that sample was originally heated can be measured.

3 Water containing the lighter ^{16}O isotope evaporates faster into and condenses slower from the atmosphere than water containing the heavier ^{18}O isotope. Therefore, there is preferential transfer of $H_2^{16}O$ from oceans to the atmosphere leaving the ocean with more $H_2^{18}O$. When water vapour reaches the polar regions and falls as rain or snow it becomes trapped in the ice and so there is more $H_2^{16}O$ in the ice compared to $H_2^{18}O$.

As a result, growing ice caps cause ocean water to become relatively richer in $H_2{}^{18}O$. When the ice sheets melt, ^{16}O-rich water is released back into the oceans, thereby decreasing the proportion of ^{18}O in the ocean's water. As a rough guide therefore, during periods of glaciation the calcite deposited in stalagmites or in corals has a higher proportion of the ^{18}O isotope whilst lower levels of ^{18}O in shells of fossil coral and cave mineral formations are associated with higher global temperatures. This is somewhat of an over-simplification, particularly with regards living organisms, as there are other factors to be considered.

4 Analysis of Neanderthal DNA suggests that there was interbreeding between Eurasian Neanderthals and colonising modern humans around 50,000–60,000 years ago and Europeans have ~2% Neanderthal DNA in their genomes. The genomes of modern humans indigenous to Africa have a very much smaller percentage of Neanderthal DNA due to back-migration of Europeans to Africa.

5 The ^{86}Sr isotope is used as it has a similar abundance of 9.9% to that of 7% for ^{87}Sr.

6 Stonehenge chronology has been determined through radiocarbon dating of antler pick tools and cremation burial remains found during excavations.

7 Bluestones is a generic term used by early excavators at Stonehenge for rocks considered exotic to the Wiltshire landscape.

8 Above-ground values. Weights are estimated from surface area and volume measurements and assuming sarsens have a density of $2.4 \times 10^3 \, \mathrm{kg/m^3}$.

FURTHER READING

The books and websites given below predominantly focus on the geological aspects of the chapter, whereas the academic articles provide more background information and in-depth analysis of the archaeological data in Section 3.7.

BOOKS

Aitken, M. J. (1990) *Science-based Dating in Archaeology*. London: Longman.

Burnham, A. (ed.) (2018) *The Old Stones: A Field Guide to the Megalithic Sites of Britain and Ireland*. London: Watkins.

Dinnis, R. and Stringer, C. (2019) *Britain: One Million Years of the Human Story*. London: Natural History Museum.

Hamilton, W. R., Woolley, A. R. and Bishop, A. C. (2013) *The Hamlyn Guide to Minerals, Rocks and Fossils*. London: Hamlyn.

Jackson, P. N. W. (2019) *Introducing Palaeontology, A Guide to Ancient Life*. 2nd Ed. Edinburgh: Dunedin Academic Press.

Park, G. (2019) *Introducing Geology, A Guide to the World of Rocks*. 3rd Ed. Edinburgh: Dunedin Academic Press.

ARTICLES

Clarke, A. J. I., Kirkland, C. L., Bevins, R. E., Pearce, N. J. G., Glorie, S. and Ixer, R. A. (2024) A Scottish Provenance for the Altar Stone of Stonehenge. *Nature*. [Online] 632. pp. 570–575. Available from: https://doi.org/10.1038/s41586-024-07652-1 [Accessed: 14th August 2024].

Evans, J. A., Chenery, C. A. and Fitzpatrick, A. P. (2006) Bronze Age Childhood Migration of Individuals Near Stonehenge, Revealed by Strontium and Oxygen Isotope Tooth Enamel Analysis. *Archaeometry*. [Online] 48 (2). pp. 309–321. Available from: https://doi.org/10.1111/j.1475-4754.2006.00258.x. [Accessed: 26th December 2023].

Green, H. S., Stringer, C. B., Collcutt, S. N., Currant, A. P., Huxtable, J., Schwarcz, H. P., Debenham, N., Embleton, C., Bull, P., Molleson, T. I. and Bevins, R. E. (1981) Pontnewydd Cave in Wales – A New Middle Pleistocene Hominid Site. *Nature*. [Online] 294. pp. 707–713. Available from: https://doi.org/10.1038/294707a0. [Accessed: 12th August 2022].

Madgwick, R., Lamb, A. L., Sloane, H., Nederbragt, A. J., Albarella, U., Parker Pearson, M. and Evans, J. A. (2019) Multi-isotope Analysis Reveals that Feasts in the Stonehenge Environs and Across Wessex Drew People and Animals from Throughout Britain. *Science Advances*. [Online] 5 (3). Article no. eaau6078. Available from: https://www.science.org/doi/10.1126/sciadv.aau6078. [Accessed: 21st January 2025].

Nash, D. J., Ciborowski, T. J. R., Ullyott, J. S., Parker Pearson, M., Darvill, T., Greaney, S., Maniatis, G. and Whitaker, K. A. (2020) Origins of the Sarsen Megaliths at Stonehenge. *Science Advances*. [Online] 6. Article no. eabc0133. Available from: https://www.science.org/doi/10.1126/sciadv.abc0133. [Accessed: 17th August 2022].

Parker Pearson, M., Bevins, R., Ixer, R., Pollard, J., Richards, C., Welham, K., Chan, B., Edinborough, K., Hamilton, D., Macphail, R., Schlee, D., Schwenninger, J.-L., Simmons, E. and Smith, M. (2015) Craig Rhos-y-felin: A Welsh Bluestone Megalith Quarry for Stonehenge. *Antiquity*. [Online] 89 (348). pp. 1331–1352. Available from: https://doi.org/10.15184/aqy.2015.177. [Date accessed: 14th August 2022].

Parker Pearson, M., Pollard, J., Richards, C., Welham, K., Casswell, C., French, C., Schlee, D., Shaw, D., Simmons, E., Stanford, A., Bevins, R. and Ixer, R. (2019) Megalith Quarries for Stonehenge's Bluestones. *Antiquity*. [Online] 93 (367). pp. 45–62. Available from: https://doi.org/10.15184/aqy.2018.111. [Accessed: 5th January 2024].

Wendt, K. A., Li, X. and Edwards, R. L. (2021) Uranium-Thorium Dating of Speleothems. *Elements*. [Online] 17 (2). pp. 87–92. Available from: https://doi.org/10.2138/gselements.17.2.87. [Accessed: 3rd December 2023].

WEBSITES

British Geological Survey. (2024) [Online] Available from: https://www.bgs.ac.uk/. [Accessed: 19th December 2024].

London Pavement Geology. (2025) [Online] Available from: https://www.londonpavementgeology.co.uk/. [Accessed 14th January 2025].

The Open University. (2024) *An Introduction to Geology*. [Online] Available from: https://www.open.edu/openlearn/science-maths-technology/an-introduction-geology/content-section-overview?active-tab=description-tab. [Accessed 27th October 2023].

The Open University. (2024) *An Introduction to Minerals and Rocks under the Microscope*. [Online] Available from: https://www.open.edu/openlearn/science-maths-technology/an-introduction-minerals-and-rocks-under-the-microscope/content-section-0?active-tab=description-tab. [Accessed: 29th October 2023].

4 Chemistry of Colour

4.1 INTRODUCTION

Colour is light that can be seen by the human eye and is generated when white light, produced by the sun or a lightbulb, interacts with materials. Depending on the material this interaction can involve light being taken in (absorption), given off (emission), passing through (transmission), bouncing off (reflection) or changing its speed and direction (refraction). Rainbows, for example, are largely the result of a refraction process and often form when the sun comes out after it has rained. Under such conditions the air is filled with raindrops and as sunlight enters the higher density medium of a water droplet it slows down and changes direction. The water droplets act as a prism and, if the sunlight passes through the raindrop at just the right angle, the light is split into its seven constituent colours – red, orange, yellow, green, blue, indigo and violet. Another eye-catching natural phenomenon is iridescence, the flash of bright colours glimpsed as some birds and insects fly past, and results from light interacting with microscopic structured surfaces of biological tissues. The final part of the chapter takes a closer look at structural colour and the green/blue iridescence of the *Calopteryx* species of dragonfly. However, in many cases colour is produced through the interaction and absorption of light with chemical compounds. It is these substances, and primarily those which are the dyes and pigments that colour fabrics or are colourants in the art world, that comprise the major part of this chapter. Vitamin A has long been associated with vision and the role of this compound in the visual pathway is considered in Section 4.2.3.

4.2 'SEEING' COLOUR

4.2.1 ABSORPTION OF LIGHT, EXCITED ELECTRONS AND MOLECULAR ORBITALS

A chemical compound gets its colour by bonding electrons absorbing energy and becoming excited. This excitation requires a specific amount of energy which is obtained from corresponding wavelengths of light and what is seen by the human eye is the complementary colour to that absorbed. Light is a form of energy and the relationship between energy and wavelength is given by the equation in Figure 4.1a. The small segment of the electromagnetic spectrum that is visible light is indicated in Figure 4.1b.

Why light of a particular frequency or wavelength is absorbed or emitted depends on the structure and bonding of the chemical compound. Understanding how this happens, whether for a coloured organic molecule (see later part of this section) or a transition metal compound (Section 4.2.2), involves taking a closer look at chemical bonding which starts with the electronic structure of the atom. In the quantum mechanical model of the atom electrons are in orbitals, regions within which there is a high probability of being found.[1] Orbitals can thus be visualised as electron clouds

DOI: 10.1201/9781003562313-4

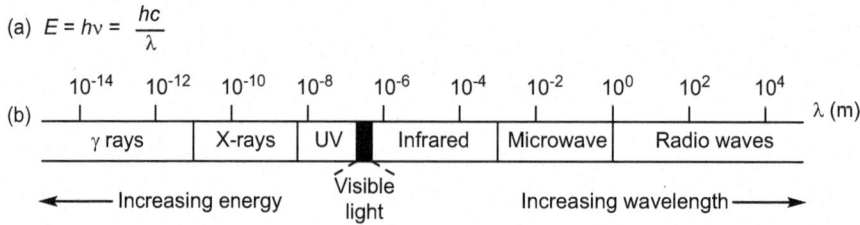

FIGURE 4.1 (a) Expression relating energy to wavelength. This expression combines Planck's equation $E = hv$ with the equation for the speed of light $c = \lambda v$, where h is Planck's constant (6.63×10^{-34} J s), c is the speed of light (3.00×10^8 m/s), λ (lambda) is the wavelength and v (nu) is the frequency. Units of frequency are Hz or s^{-1} and wavelengths are given in metres (m) or more often nm (nanometre, 10^{-9} m; e.g. 650 nm = 6.50×10^{-7} m). The spectrum of visible light ranges from red at the long wavelength end (750 nm) to violet at the short wavelength end (400 nm). The inverse relationship between energy and wavelength means that violet light has the most energy and red the least. (b) Full range of electromagnetic radiation from the high energy and short wavelengths of γ-rays to the low energy and long wavelengths of radio waves.

surrounding the nucleus and they have characteristic shapes and energies (Figure 4.2) which are a result of the mathematics used. Superimposing all of the orbitals of an atom gives roughly a spherical shape and thus justifies the usual representation of an atom as a sphere.

The way the electrons of an atom are distributed amongst the various orbitals is the electron configuration. Orbitals are filled in order of increasing energy with no more than two electrons per orbital in accordance with the rules of quantum mechanics. The single electron in a hydrogen atom occupies the lowest energy $1s$ orbital and the electron configuration is $1s^1$. The order of orbital energies is $1s < 2s < 2p < 3s < 3p < 4s < 3d < 4p < 5s$ etc. and this theoretical sequence is verified experimentally from ionisation data and atomic emission spectra. The electron configuration of potassium, for example, is $1s^2\,2s^2\,2p^6\,3s^2\,3p^6\,4s^1$ where the superscripts indicate the number of electrons in the various types of orbital.

Electrons in atomic orbitals can be described by wave functions in quantum mechanics theory and when wave functions for two orbitals interact this can be in a constructive or destructive manner resulting in two molecular orbitals. Constructive interaction in which the wave functions are added together leads to electron density being concentrated in the region between the two nuclei. This is the bonding molecular orbital and holds the atoms together in a covalent bond. Covalent bonding is often described more simply as the overlap of two atomic orbitals in which parts of the orbitals occupy the same space with the two electrons aligning between the two nuclei. As electrons in this molecular orbital are attracted to both nuclei they are more stable (have lower energy) than in the original atomic orbitals. The other molecular orbital, constructed by subtracting the wave functions, more or less cancels out electron density in the internuclear region which tends to push the nuclei apart. This type of orbital is an antibonding molecular orbital and is higher than the energy of the atomic orbitals (Figure 4.3).

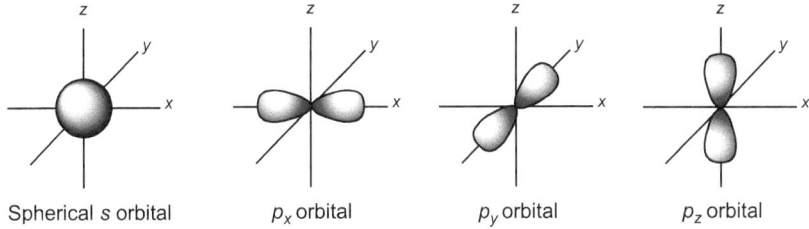

| Spherical s orbital | p_x orbital | p_y orbital | p_z orbital |

Three equivalent p orbitals, each having two lobes

FIGURE 4.2 Atomic orbitals, energies, shells and sub-shells. The relative energies and types of orbital are identified by quantum numbers. The principal quantum number (n) has values 1, 2, 3 ... and denotes the size and overall energy of the orbital. As n increases the orbital becomes larger and has higher energy. Negatively charged electrons are attracted to the positively charged nucleus by electrostatic forces. However, electrons further from the nucleus are less strongly bound and so have higher (potential) energy. The angular momentum quantum number (l) defines the shape of the orbital designated by the letters, s, p, d and f, and the magnetic quantum number (m) the orientation of the orbital. An s orbital is shaped like a sphere and a $2s$ orbital has a larger radius than a $1s$ orbital. The p orbitals resemble dumb-bells with two lobes of electron density (indicated by shaded regions in the diagram). There are three p orbitals (labelled p_x, p_y and p_z according to the axis along which they lie) for each principal quantum number and they are orientated at 90° to each other and are of equal energy. The d orbitals are discussed in Section 4.2.2 whilst the f orbitals are difficult to visualise and will not be considered further. The collection of orbitals with the same value of n is called an electron shell and thus the $2s$ and $2p$ orbitals are in the second shell with the $2s$ and $2p$ orbitals being referred to as sub-shells.

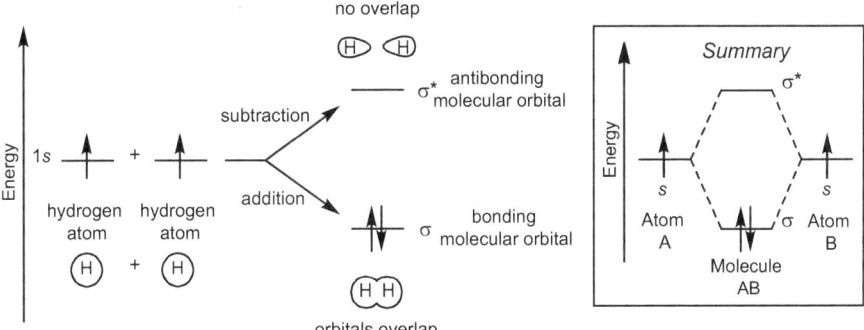

FIGURE 4.3 Energy level diagram for formation of a sigma (σ) covalent bond in H_2. The combination of the two hydrogen ($1s$) atomic orbitals forms a new bonding and antibonding pair of molecular orbitals, designated as σ and σ^* (sigma star) respectively. The small arrows represent electrons and the two arrows in the bonding molecular orbital, as in a doubly occupied atomic orbital, point in opposite directions to represent opposite electron spins (one of the laws of quantum mechanics). As the two electrons in H_2 occupy the bonding molecular orbital which is lower in energy than the $1s$ atomic orbitals, the H_2 molecule is more stable than the two separate hydrogen atoms. Inset is the usual representation of covalent bonding energy level diagrams; the levels in the centre represent the molecular orbitals formed by the interaction of the atomic orbitals (represented by the energy levels on the left and right) of the atoms forming the molecule.

In a carbon-carbon double bond such as in ethene, $H_2C=CH_2$, one bond is produced by end-to-end orbital overlap between the adjacent carbon atoms to form sigma (σ) bonding and antibonding (σ*) molecular orbitals. The other bond is produced by sideways overlap of the lobes of the two p orbitals to give pi (π) bonding and antibonding (π*) molecular orbitals. The π bonding molecular orbital has two parts, one on either side of the σ orbital (Figure 4.4a). The electron density in the π bonding orbital is no longer directly between the two nuclei, with the result that it is not held as strongly as in a σ bonding orbital.

A bonding electron can absorb light energy and be promoted from a bonding molecular orbital to an unoccupied molecular orbital and this excited state is less stable than the original ground state. Promotion of an electron from the highest occupied molecular orbital (HOMO) to the lowest unoccupied molecular orbital (LUMO) of an alkene (Figure 4.5a) requires the smallest amount of energy of the transitions which are possible. This excitation can be induced by applying light to the molecule but only if the wavelength employed corresponds exactly to the amount of energy separating the HOMO and LUMO orbitals. An electron in ethene can be promoted from the π to the π* orbital by light with a wavelength of 165 nm which is in the ultraviolet region of the electromagnetic spectrum. Using the expression in Figure 4.1 this equates to 726 kJ/mol of energy. Higher energy, shorter wavelength light (less than 150 nm) is necessary to excite a σ electron as there is a larger energy gap between the bonding and antibonding orbitals.

In conjugated systems where the molecule has alternating single and double bonds such as butadiene, $H_2C=CH-CH=CH_2$, the π orbitals from each double bond interact to form a new set of bonding and antibonding orbitals (Figure 4.5b). The difference in energy between the ground and excited states for the $\pi_2 \rightarrow \pi_3$* transition in butadiene is less than the $\pi \rightarrow \pi$* transition in ethene and consequently butadiene absorbs energy at the much longer wavelength (lower energy) of 217 nm. Additional conjugation further decreases the HOMO-LUMO energy gap which pushes the position of the lowest energy absorption to longer wavelength. Eventually the HOMO-LUMO gap becomes small enough so that the absorption enters the visible range and highly conjugated polyenes are coloured. β-Carotene (Figure 4.5c) found in carrots, for example, absorbs light at wavelengths 453 and 483 nm in the blue region of the visible spectrum so the complementary, and therefore the observed, colour is orange.

4.2.2 Transition Metal Compounds, Geometry and d Orbital Splitting

Transition metal compounds produce colour in an analogous manner to conjugated molecules with the light emerging from a substance being the original light minus the absorbed wavelengths which have caused the excitation of an electron between two energy levels. Before considering how an energy gap is generated between two sets of orbitals in these inorganic compounds, it is first necessary to consider some fundamental aspects of their chemistry.

Transition metals are d-block elements characterised by the number of electrons in d orbitals; iron, for example, has six d electrons and an electronic configuration of $1s^2\ 2s^2\ 2p^6\ 3s^2\ 3p^6\ 3d^6\ 4s^2$. This can be condensed to [Ar] $3d^6\ 4s^2$ to focus on the outermost electrons in the atom; [Ar] is the electronic configuration of argon, the

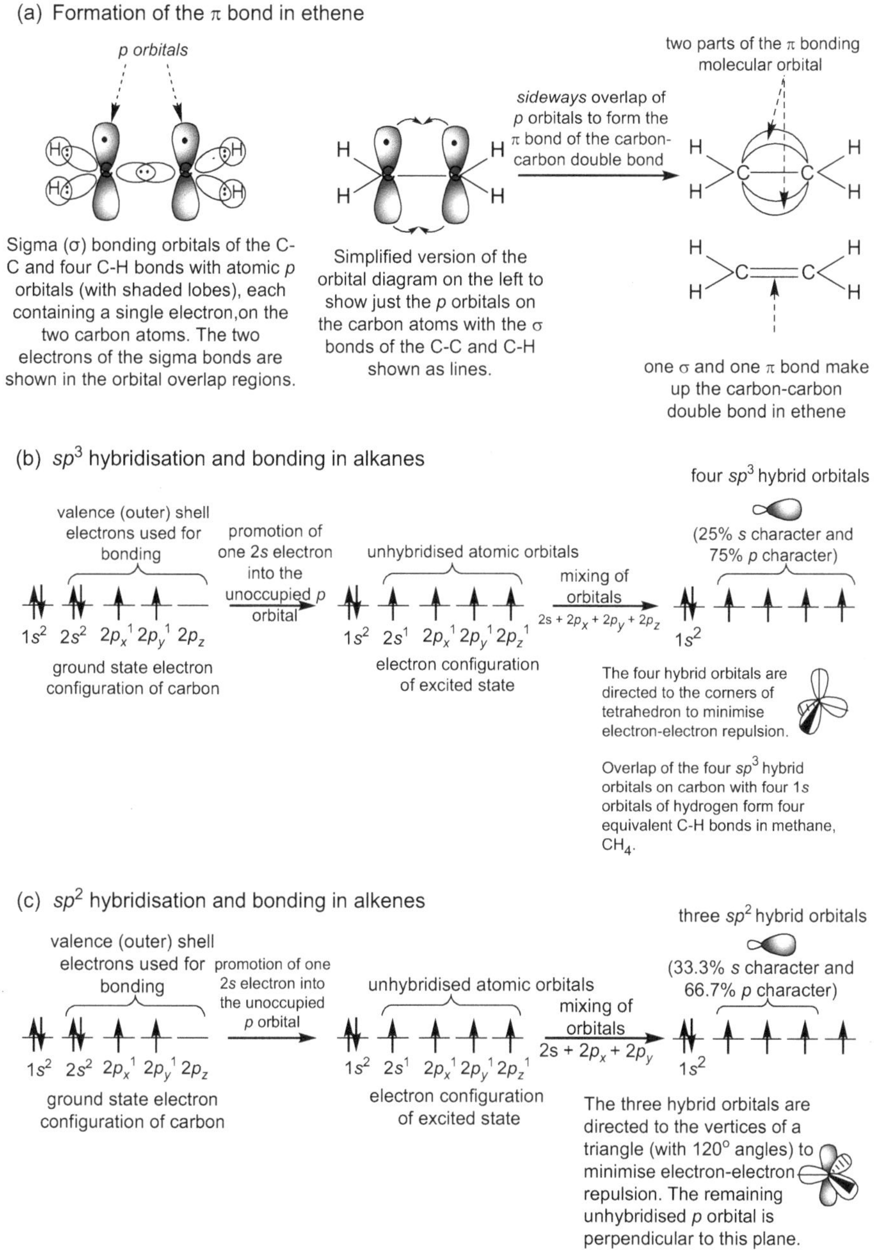

(a) Formation of the π bond in ethene

p orbitals

Sigma (σ) bonding orbitals of the C-C and four C-H bonds with atomic *p* orbitals (with shaded lobes), each containing a single electron, on the two carbon atoms. The two electrons of the sigma bonds are shown in the orbital overlap regions.

Simplified version of the orbital diagram on the left to show just the *p* orbitals on the carbon atoms with the σ bonds of the C-C and C-H shown as lines.

sideways overlap of *p* orbitals to form the π bond of the carbon-carbon double bond

two parts of the π bonding molecular orbital

one σ and one π bond make up the carbon-carbon double bond in ethene

(b) *sp*³ hybridisation and bonding in alkanes

four *sp*³ hybrid orbitals

valence (outer) shell electrons used for bonding

promotion of one 2s electron into the unoccupied *p* orbital

unhybridised atomic orbitals

mixing of orbitals
$2s + 2p_x + 2p_y + 2p_z$

(25% *s* character and 75% *p* character)

$1s^2 \ 2s^2 \ 2p_x^{\ 1} 2p_y^{\ 1} 2p_z$

$1s^2 \ 2s^1 \ 2p_x^{\ 1} 2p_y^{\ 1} 2p_z^{\ 1}$

$1s^2$

ground state electron configuration of carbon

electron configuration of excited state

The four hybrid orbitals are directed to the corners of tetrahedron to minimise electron-electron repulsion.

Overlap of the four *sp*³ hybrid orbitals on carbon with four 1s orbitals of hydrogen form four equivalent C-H bonds in methane, CH_4.

(c) *sp*² hybridisation and bonding in alkenes

three *sp*² hybrid orbitals

valence (outer) shell electrons used for bonding

promotion of one 2s electron into the unoccupied *p* orbital

unhybridised atomic orbitals

mixing of orbitals
$2s + 2p_x + 2p_y$

(33.3% *s* character and 66.7% *p* character)

$1s^2 \ 2s^2 \ 2p_x^{\ 1} 2p_y^{\ 1} 2p_z$

$1s^2 \ 2s^1 \ 2p_x^{\ 1} 2p_y^{\ 1} 2p_z^{\ 1}$

$1s^2$

ground state electron configuration of carbon

electron configuration of excited state

The three hybrid orbitals are directed to the vertices of a triangle (with 120° angles) to minimise electron-electron repulsion. The remaining unhybridised *p* orbital is perpendicular to this plane.

FIGURE 4.4 Bonding in alkanes and alkenes. (a) Overlap of *p* orbitals to form the π bond in ethene. The electron in each of the *p* orbitals is shown in the upper lobes but they can equally be found in the lower lobes (another law of quantum mechanics) which results in the electron density of the π bond being both above and below the plane of the C–C σ bond. The diagram in the middle shows how the sideways overlap of *p* orbitals forms the π carbon-carbon bond. The diagram on the left is an oversimplification of the σ-bonding framework in ethene as

(Continued)

FIGURE 4.4 (*Continued*) formation of the C–C and C–H bonds involves hybrid orbitals. (b) Formation of sp^3 hybrid orbitals and bonding in methane, CH_4. A carbon atom has the electron configuration $1s^2\,2s^2\,2p^2$ or $1s^2\,2s^2\,2p_x^1\,2p_y^1$, showing that two of the three $2p$ orbitals contain one electron. There are four valence electrons: the two in the $2p_x$ and $2p_y$ orbitals and the doubly occupied $2s$ orbital. However, for the carbon atom to form four bonds there must be four singly occupied valence orbitals and so the carbon atom assumes the more energetic configuration $1s^2\,2s^1\,2p_x^1\,2p_y^1\,2p_z^1$ and recoups the energy required to do this by forming bonds to hydrogen atoms. To form the four equivalent bonds in CH_4 the $2s$, $2p_x$, $2p_y$ and $2p_z$ orbitals, each of which contains one electron, are hybridised (or mixed) to form four equivalent sp^3 hybrid orbitals which overlap with the $1s$ atomic orbitals on hydrogen to form the four σ bonding molecular orbitals. (c) Formation of sp^2 hybrid orbitals in ethene and the σ bonding framework in ethene. The s orbital and the $2p_x$ and $2p_y$ orbitals combine to give three sp^2 orbitals which point to the corners of an equilateral triangle and the third p orbital is unhybridised and is used to form the π bond. The three hybrid sp^2 atomic orbitals (each of which contains one electron) form the σ molecular orbitals of the C–H and C–C bonds; overlap of sp^2–sp^2 gives the C–C σ bond and sp^2–$1s$ overlap gives the C–H σ bonds.

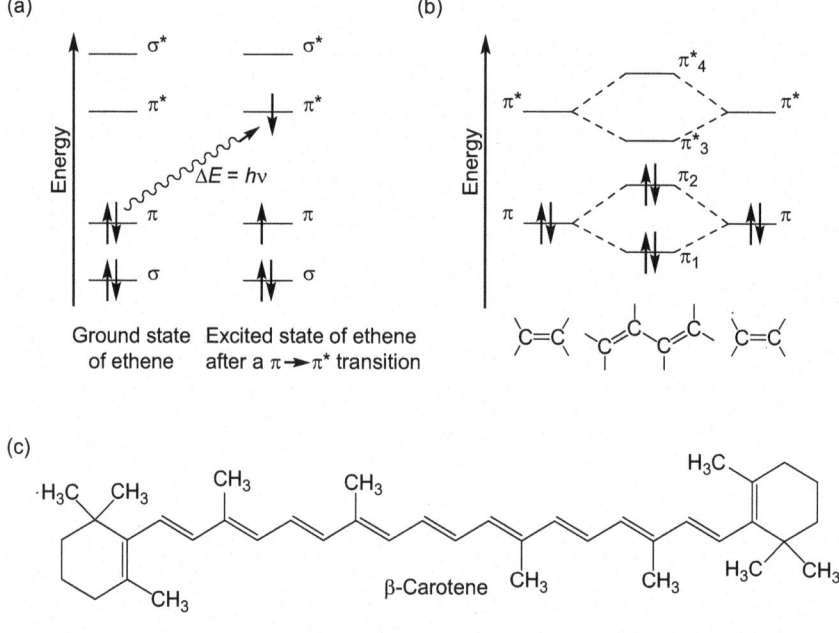

FIGURE 4.5 (a) Excitation of a π electron in ethene. (b) Orbital diagram for butadiene. (c) Structure of β-carotene.

nearest noble gas element. When transition metals are oxidised to form cations they lose the $4s$ electrons before they lose electrons from the d orbitals and so Fe^{2+} has an electron configuration of $[Ar]\,3d^6$. Transition metal ions form coordination compounds in which the central metal ion is bonded to molecules or ions called ligands. For example, potassium hexacyanoferrate(II) (commonly known as potassium ferrocyanide) has the formula $K_4[Fe(CN)_6]$ and is composed of a central iron(II) ion, Fe^{2+}, coordinated to six cyanide ions (CN^-). The resulting complex ion, $[Fe(CN)_6]^{4-}$, has a

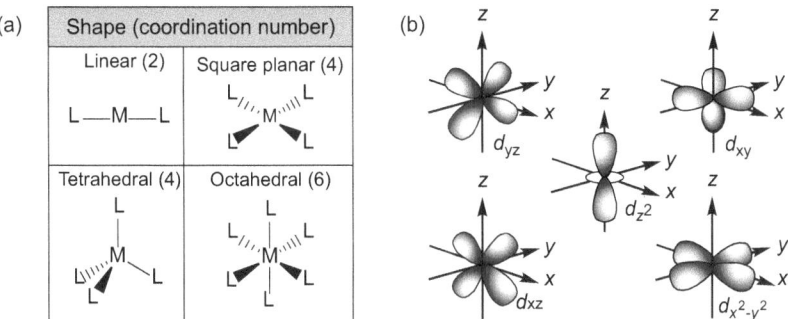

FIGURE 4.6 (a) Shapes and coordination numbers of the four most common types of transition metal complexes; M = central metal ion and L = ligand. (b) Shapes of the five d orbitals as defined by their x, y and z coordinates. Four (d_{xy}, d_{xz}, d_{yz} and $d_{x^2-y^2}$) of these orbitals have a cloverleaf pattern with four lobes of electron density and each orbital lies primarily in a plane. The d_{xy}, d_{xz}, d_{yz} orbitals lie in planes defined by the xy-, xz- and yz-axes respectively with the lobes oriented between the axes. The lobes of the $d_{x^2-y^2}$ orbital also lie in the xy plane but the lobes lie along the x- and y-axes. The fifth orbital $\left(d_{z^2}\right)$ has two lobes oriented along the z-axis and a doughnut-shaped ring in the xy plane.

negative charge which is balanced by an equal and opposite charge, in this instance four potassium ions, K^+. Complexes can also be positively charged with a negative counterion, for example, hydrated iron(III) chloride, $[Fe(H_2O)_6]Cl_3$, or neutral such as in cisplatin, *cis*-$[PtCl_2(NH_3)_2]$.[2] Ligands can be either neutral molecules (such as H_2O and NH_3) or anions (such as chloride, Cl^-, and cyanide, CN^-). The characteristic feature of ligands is that they contain a lone pair of electrons which is donated to empty orbitals on the metal ion (the acceptor) to form metal-ligand coordinate covalent bonds.

The number of atoms directly bonded to the metal ion in a complex is known as the coordination number and the specific geometric arrangement of ligands around the central metal ion is the shape. Complexes can exhibit a variety of coordination numbers and shapes; attaching two ligands on opposite sides of a metal ion results in a linear shape, whereas having four ligands can give rise to tetrahedral or square planar geometry. The most common coordination number is six with the ligands in an octahedral arrangement around the central metal ion (Figure 4.6a).

Metal-ligand interactions in complexes are in essence covalent bonds which can be described in terms of molecular orbitals formed by atomic orbital overlap between metal and ligands. The resulting molecular energy level diagrams are complex but generate HOMO and LUMO orbitals separated by an energy gap which can account for the colour of transition metal compounds. However, an alternative bonding model using ligand field theory is a more pragmatic approach to rationalising the colour of transition metal complexes. In ligand field theory the ligands are considered to be points of negative charge and are attracted electrostatically towards the positively charged metal ion. Whilst the ligand electrons are attracted to the metal ion they are also repelled by the electrons in the d orbitals of the metal ion. The magnitude of this effect is determined by the juxtaposition of electron density in the five d orbitals and the ligand geometry around the central metal ion.

4.2.2.1   Octahedral Complexes, $[ML_6]^{n+}$

Six ligands approaching a central metal ion along the x-, y- and z-axes create an octahedral field. As ligands are regions of negative charge, the d orbitals on the metal ion which point along the x-, y- and z-axes will be in a region of higher negative field than those which are oriented between the axes. This difference in orientation means that electrons in the metal ion's $d_{x^2-y^2}$ and d_{z^2} orbitals experience a stronger repulsion than those in the d_{xy}, d_{xz} and d_{yz} orbitals and as a result the degeneracy of d orbitals is broken (Figure 4.7a). The energy gap between the two sets of d orbitals is called the ligand field splitting energy and given the symbol Δ_o, where the subscript indicates octahedral.

4.2.2.2   Tetrahedral Complexes, $[ML_4]^{n+}$

A tetrahedral arrangement of four negative charges around a metal ion creates an electrostatic field which causes a d orbital spitting pattern which is the reverse of that observed for octahedral complexes. The three d_{xy}, d_{xz} and d_{yz} orbitals are higher in energy, whereas the $d_{x^2-y^2}$ and d_{z^2} orbitals are below them in energy and, as there are only four ligands in a tetrahedral complex, the ligand field splitting energy (Δ_t) is smaller than that in an octahedral complex (Figure 4.7b).

The gap between the d_{xy}, d_{xz} and d_{yz} orbitals and the $d_{x^2-y^2}$ and d_{z^2} set is of the same order of magnitude as the energy of a photon of visible light. Using Planck's equation (see Figure 4.1a) the absorption of visible light with wavelength $\lambda = (h\,c)/\Delta$ can cause the excitation of an electron from a lower-energy d orbital to a higher-energy d orbital which is called a d-d transition. The Cr^{3+} ion has three d electrons, each singly occupying three d orbitals, and when light is absorbed by an octahedral complex such as $[Cr(H_2O)_6]^{3+}$ an electron is promoted from a t_{2g} orbital to a vacant higher

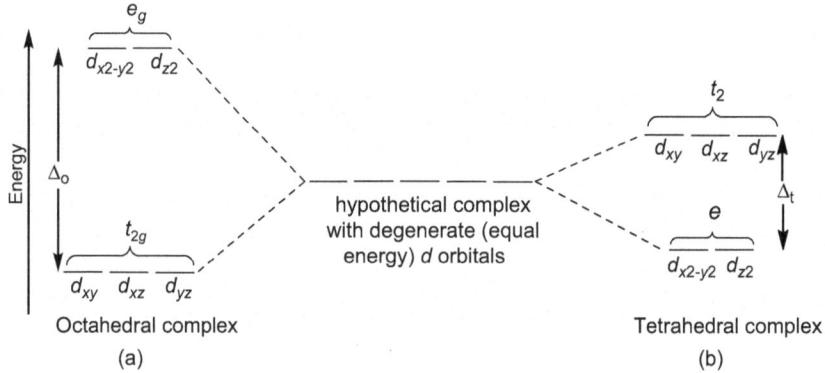

FIGURE 4.7   Energy level diagrams showing the splitting of a set of d orbitals by octahedral and tetrahedral electrostatic ligand fields. The t_{2g}, e_g, t_2 and e are symmetry labels arising from group theory. Similar reasoning to that described in the text gives energy diagrams for square planar, linear and other shapes of complexes.

energy e_g orbital (Figure 4.8a). The magnitude of the ligand field splitting energy, and consequently the colour of the complex, depends on the nature of the metal ion, the charge on the ion and the ligands. For example, the $[Cr(H_2O)_6]^{3+}$ ion is violet whereas $[Cu(H_2O)_6]^{2+}$ is blue, $[Co(H_2O)_6]^{2+}$ pink and $[CuCl_4]^{2-}$ yellow. A greater charge on the metal ion draws the ligands closer causing a larger repulsion with the d orbitals which increases the magnitude of the energy gap between the orbitals. For example, iron(II) sulphate, $[Fe(H_2O)_6] SO_4$, is pale green as it absorbs lower energy wavelengths whilst iron(III) chloride, $[Fe(H_2O)_6] Cl_3$, is yellow-orange as the absorption moves towards the blue (and higher energy end) of the visible spectrum.

Many transition metal compounds are coloured because of d-d transitions. However, other compounds, such as potassium manganate(VII) ($KMnO_4$) and potassium dichromate(VI) ($K_2Cr_2O_7$), derive their colours from a different type of excitation involving d orbitals. The purple manganate $\left(MnO_4^-\right)$ and orange dichromate $\left(Cr_2O_7^{2-}\right)$ ions are complexes in which the metals have d^0 electronic configurations and so the absorbance peaks in their visible spectra cannot be due to d-d transitions. The colour of these complexes, which is often very intense, is due to a charge-transfer transition in which an electron on one of the oxygen ligands is excited into a vacant d orbital on the metal ion (Figure 4.8b). Charge-transfer transitions are also responsible for the colours of the artists' pigments chrome yellow ($PbCrO_4$) and cadmium yellow (CdS) which have d^0 and d^{10} configurations respectively. The intensely coloured pigment Prussian blue, $Fe^{III}_4[Fe^{II}(CN)_6]_3$, owes its colour to a metal-to-metal charge-transfer process which reverses the oxidation states of the two metal sites.

Ligand field effects caused by small amounts of Cr^{3+} impurities in ruby and emerald gemstones are responsible for their red and green colours respectively. Rubies and emeralds are crystals of aluminium oxide (Al_2O_3) and beryllium aluminium silicate $[Be_3Al_2(SiO_3)_6]$ respectively in which about 1% of the Al^{3+} ions have been replaced by Cr^{3+} ions. The ligands surrounding the embedded Cr^{3+} ions create a field effect which causes d orbital splitting. The ligand field splitting is greater in ruby than in emerald and the absorption peaks in the visible spectrum of ruby are at shorter

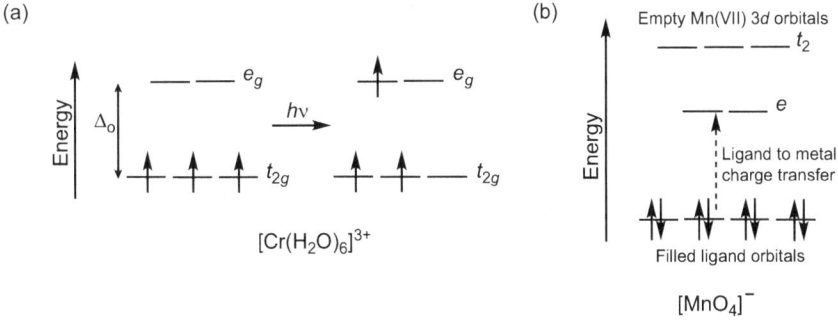

FIGURE 4.8 Purple colours in transition metal compounds due to absorbance in the yellow/green portion of the visible spectrum. (a) A d-d transition of an octahedral Cr^{3+} complex such as $[Cr(H_2O)_6]^{3+}$ and (b) ligand-to-metal charge transition in the MnO_4^- ion.

wavelengths, i.e. higher energy, compared to those in emerald, thus resulting in different colours of light being transmitted by the two gemstones.

4.2.3 PHOTORECEPTORS, ISOMERISATION AND A CHEMICAL CASCADE

Light reflected by an object is sensed by the eye and converted into a neural signal which is transmitted to the brain. Rods and cones, specialised cells at the back of the retina, contain light-sensitive pigments that absorb photons reflected by the object and begin the process of creating the neural signal. Three types of cone cells, each containing a photoreceptor, are responsible for colour vision whilst a fourth photoreceptor found in rod cells mediates vision in dim light conditions. All four of these photoreceptors comprise a seven-transmembrane domain protein, opsin, a member of the G protein-coupled receptor (GPCR) superfamily, covalently bound to the chromophore 11-*cis*-retinal. Differences in the amino acid environment of the opsin proteins surrounding the binding site of the chromophore alter specific steric and electrostatic interactions, thus affecting the range of wavelengths which are absorbed. As a result, the absorption spectra between the three different cone photoreceptors vary, with maximum absorbances of 420, 534 and 564 nm corresponding to blue, green and red light respectively (Figure 4.9). Human colour vision is

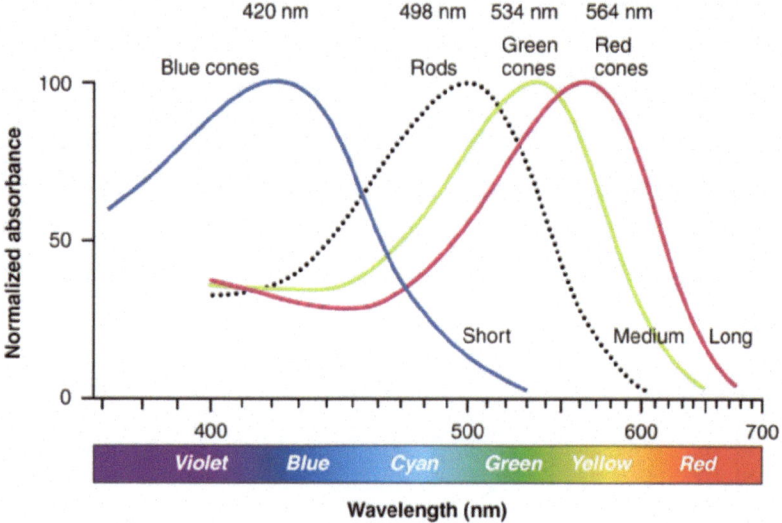

FIGURE 4.9 Spectral sensitivity of cone and rod photoreceptors. Cone photoreceptors maximally sensitive to blue, green and red light are more appropriately referred to as short (S), medium (M) and long (L) wavelength respectively. All the cones are sensitive to a broad band of wavelengths and the maximum absorbance of the long wavelength cone at 564 nm does not correspond to red light but to yellowish green light. At 564 nm the green cone is also strongly stimulated and when both red and green cones are strongly stimulated the brain perceives a yellowish colour. All photoreceptors are capable of absorbing light of almost any wavelength and the peak maximum just indicates the particular wavelength of light that the photoreceptor absorbs most readily. This Photo by Unknown Author is licensed under CC BY.

trichromatic as the blue, green and red light can be mixed to make any other colour. The colour perceived by individuals is a result of the joint action and specific ratio of activity of the three cone photoreceptors.

Activation of cone photoreceptors requires bright light and cone cells are concentrated opposite the centre of the pupil, the opening towards the front of the eye formed by the iris. Rod cells are more numerous in the periphery of the retina with a lower density towards the centre but are more sensitive to light and can generate a response from just a single photon. Under low light levels the single type of rod photoreceptor (rhodopsin) gives the same response at all wavelengths and leads to low-resolution detection of shape, contrast and movement on a black-and-white scale but no colour information.

When opsin-bound 11-*cis*-retinal absorbs a photon, an electron in a π bonding orbital is promoted to an antibonding π* orbital, thus temporarily destroying the double bond character between carbons 11 and 12. In this excited state, rotation about the single carbon-carbon bond occurs rapidly and reformation of the π bond gives the more stable *trans*-form (Figure 4.10).

Cis-trans isomerisation alters the shape of the chromophore molecule from a bent into a linear form and *trans*-retinal is the wrong shape for the opsin binding pocket.

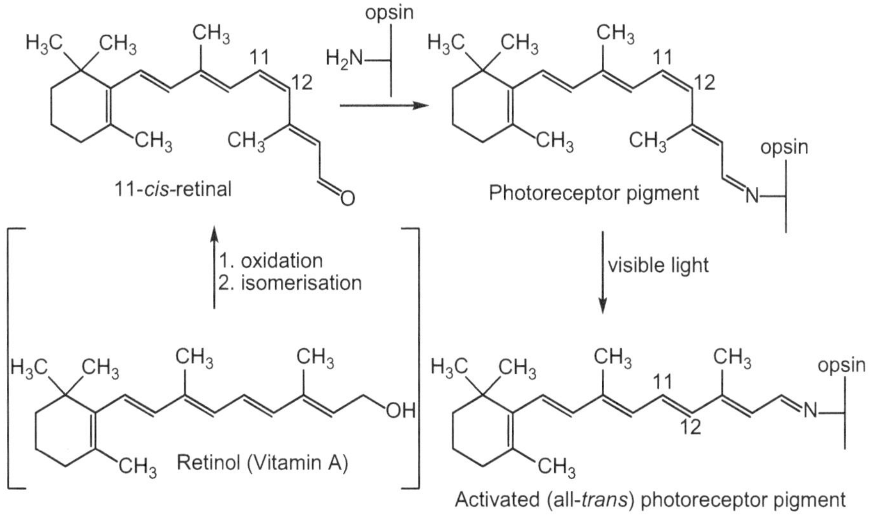

FIGURE 4.10   First steps in the process of vision. The chromophore is bound to the opsin protein via a C=N bond through a reaction between the aldehyde group of 11-*cis*-retinal and the amine on one of the lysine amino acid side chains. Absorption of light induces a *cis-trans* isomeric change in the structure of the protein-bound chromophore. Note that the precursor to 11-*cis*-retinal is the alcohol, all-*trans*-retinol, commonly known as vitamin A. This molecule cannot be synthesised in the body and has to be acquired through the diet. Vitamin A is converted to 11-*cis*-retinal in two enzymic reactions, first oxidation of the alcohol to an aldehyde and then isomerisation of the double bond between carbons 11 and 12. Vitamin A is also formed by enzymic cleavage of β-carotene in the small intestine. The structure of β-carotene is shown in Figure 4.5c.

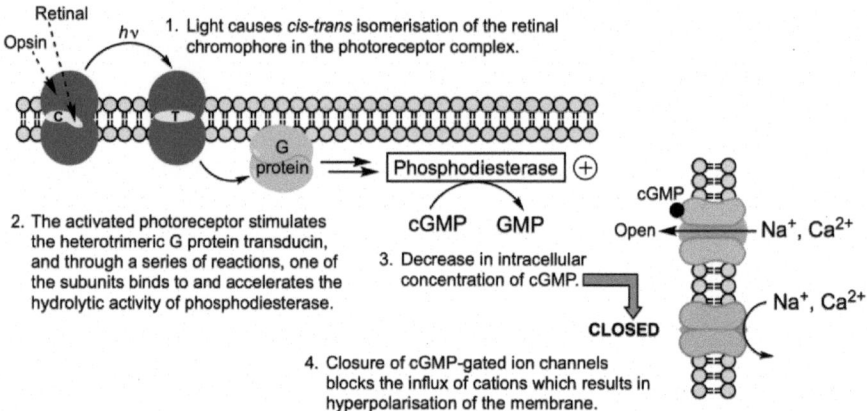

FIGURE 4.11 Process of phototransduction in rod and cone cells. Note that cGMP is the abbreviation for cyclic guanosine monophosphate and GMP is guanosine monophosphate.

This causes a conformational change in the protein component of the photopigment which allows the G protein transducin to bind (Figure 4.11). Activation of trans-ducin triggers a series of intracellular reactions, culminating in the hydrolysis of cyclic guanosine monophosphate (cGMP). The fall in cGMP concentration causes cGMP-gated sodium and calcium channels to close which leads to hyperpolarisation of the cells, i.e. the charge inside the cell membrane becomes more negative. Rods and cones are very small cells (20–30 microns) and hyperpolarisation is communi-cated to the terminals by a process of decremental conduction rather than an action potential. Hyperpolarisation causes a reduction of the release of the neurotransmitter glutamate at the photoreceptor synapse. Although inhibitory information is received at the synapse, it is the relative change in potential that enables the visual signal to be relayed through a series of other cells in the retina.[3] Finally the ganglion cells receive the signal and send action potentials along their axons (which together form the optic nerve), thereby transmitting the message to the brain which interprets it as vision.

Many biochemical reactions are required to deactivate the phototransduction cas-cade and reset the photoresponse back to the ground state. For example, trans-retinal dissociates from the opsin protein and is converted back into 11-cis-retinal in a series of enzymatic steps comprising the visual cycle. Within this cycle, trans-retinal is reduced to the alcohol trans-retinol which is converted to 11-cis-retinol and then an oxidation reaction gives 11-cis-retinal. The chromophore then covalently bonds to the opsin protein to regenerate the photoreceptor pigment.

4.2.4 PERCEPTION

Optic nerve fibres transmit the electrical signals received from the retina to neurons in the visual cortex, areas at the back of both hemispheres of the brain, where the information is processed and decoded into meaningful images. What is perceived as colour is a combination of the retinal images and information stored in the memory.

Colour therefore is a psychological phenomenon that has, from the earliest times, had practical applications in helping people to navigate their natural environment, such as finding ripe fruits amongst leaves or in a modern context following traffic light signals when traversing the road network safely.

Red (R), green (G) and blue (B) are the sets of wavelengths of light that the three cone receptors are tuned to in the human eye and the colour perceived is the complementary colour to that absorbed. A magenta colour therefore is the visual sensation when the green wavelengths have been absorbed, reflecting the red and blue wavelengths and these relationships can be readily seen in a Venn diagram (Figure 4.12a). RGB are referred to as the primary colours and cyan (C), magenta (M) and yellow (Y) as the secondary or complementary colours. The RGB system is used in digital media such as computer screens or where the red, green and blue lights are added to make other colours. Paints, pigments and inks generate colour through a subtractive mixing system as some wavelengths are absorbed, i.e. subtracted from the incoming white light. The most effective set of primary colours in the subtractive system are CMY and combining all three primaries absorbs all light and produces black (Figure 4.12b). Combining all three primary colours of light in the direct additive system creates white.

Traditional art forms considered red, yellow and blue as primary colours and this RYB model is still a useful tool for artists working in paints and pigments. This type of colour prediction does not consider other effects such as light scattering by pigment particles and skilled artists typically rely on their experience and own recipes to mix desired colours.

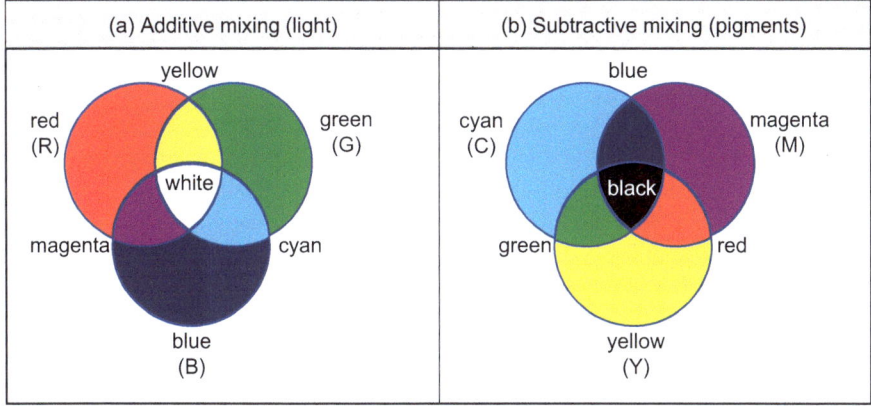

FIGURE 4.12   Additive and subtractive colour. (a) Venn diagram to show additive colour using red, green and blue light (circles) as the primary colours. The overlapping region between two circles shows the effect of adding two colours of light and the central area shows the effect of adding all three colours. (b) Venn diagram to show subtractive colour using cyan, magenta and yellow as the primary colours. Subtractive mixing of any of the two primary colours produces one of the additive primaries. Complete absorption of all three primaries results in no light being reflected and produces black.

Colour appeals to an individual's aesthetic sense and has the power to affect emotion and moods which can be used by businesses to stimulate interest in their products. The physical generation of coloured substances for practical purposes, such as dyeing cloth or painting pictures, has been carried out for centuries in workshops and laboratories of many civilisations and gives the modern world a tangible link to the past.

4.3   COLOURED COMPOUNDS

4.3.1   INTRODUCTION

Using colourants in the 21st century to dye fabrics or in paint formulations is a blend of the old and new with modern synthetic materials complementing those extracted from traditional plant, animal and mineral sources. Coloured fabrics and paints are readily available today whereas in the past some pigments were prohibitively expensive and at times only higher echelons of society were permitted to wear, by law, certain coloured textiles. Creating and using dyes and pigments are ancient skills which have evolved and adapted over the centuries to meet the requirements of the time. Inks, for example, were originally finely ground particles of soot or coloured minerals which were bound together with gum into a paste and then mixed with water immediately before being used for writing and drawing. Handwriting inks were not suitable for the mechanical printing technology of the 16th century and special oily formulations involving pigment particles suspended in turpentine and walnut oil were developed. In modern photocopying processes the inks are a blend of pigment and plastic particles contained in toner cartridges which come in four colours: black, cyan, magenta and yellow. Printing then involves a combination of light and electrostatic interactions to transfer the ink to the paper and then the action of heat melts the plastic material in the toner which fuses the pigment to the page. Current handwriting inks are predominantly solutions of dyes in water or organic solvents with other ingredients to ensure a smooth and consistent application of the ink to the paper. Coloured compounds also have a long association with the food and cosmetic industries which are highly regulated to ensure customer safety with relevant ingredients identified by E or colour index (C.I.) numbers.

Depending on whether a coloured compound is used as a dye or a pigment is fundamentally based on its solubility. Pigments are insoluble substances, which can be either inorganic or organic compounds, whereas most dyes are soluble organic molecules which are attached to fabrics through physical deposition or through the formation of bonds during a reaction carried out in solution. Paints adhere to a surface through the action of the liquid binding oil which solidifies and holds the pigment particles in place. In modern acrylic paints, for example, the double bonds in acrylic acid ($H_2C{=}CH{-}CO_2H$) polymerise through a radical-initiated reaction and the polymers interlink and form intramolecular interactions producing a rigid material. Over time paints can lose their colour as pigment compounds absorb light, particularly blue wavelengths and ultraviolet radiation. Red and green colours tend to be lost first as these absorb blue wavelengths of light and the painting appears bluish.

Textiles are organic materials and rarely survive in the archaeological record as they deteriorate quickly in most climatic and soil conditions. Some of the best examples of early coloured textiles are the wrappings of mummified bodies from ancient Egypt and the Paracas peninsula of Peru, and they survived only because of their burial in dry desert conditions away from sunlight. Fabric dyeing remained largely unchanged from the ancient methodology until 1856 and the advent of the synthetic dye industry (see Sections 4.4 and 4.3.3.4). Synthetic dyes can be specifically designed to optimise covalent bond formation with fibre molecules and the reactive triazine dyes developed in the mid-1950s proved particularly effective for colouring cellulose and some man-made textiles. Moreover, because these dyes were covalently bound to the fibre polymers the textiles did not lose their colour when they were washed. Although today most coloured textiles are dyed with synthetic compounds, either novel chemical structures or natural products prepared in the laboratory, traditional methods of colouring wool, silk and cottons with vegetable dyes have enjoyed a resurgence in recent years. Recipes for dyeing fabrics with natural dyes are readily available in books and online and several such procedures are explored in Chapter 6.

4.3.2 Pigments in Ancient and Modern Times

4.3.2.1 The Original Colour Palette

Some of the oldest surviving coloured materials are in prehistoric cave paintings. These early pigments were largely produced from powdered rocks and minerals and then mixed with a liquid binding agent such as plant sap to make the paint. Polyunsaturated fatty acids in the binder would have polymerised, thus adhering the pigment to the cave wall. Red and yellow colours were obtained from ochre, a common clay-based material coloured by iron oxides, whilst white pigments came from chalk or shells and the blacks from charcoal or soot. Ochre pigments could range in colour depending on the relative amounts of red haematite [iron(III) oxide, Fe_2O_3] and yellow limonite [hydrated iron(III) oxide-hydroxides, $FeO(OH) \cdot nH_2O$] present and the size of granules. At a later stage it was found that the more abundant yellow ochres could give a red tint when heated, now known to be the result of a dehydration reaction and formation of the red iron oxide Fe_2O_3.

4.3.2.2 Enriching the Palette

One of the first synthetic pigments was created by the ancient Egyptians who made a blue compound to mimic the colour which could be obtained from the rare and expensive rock lapis lazuli. Blue was considered a high-status colour by the Egyptians and important in images of their royalty and gods. Modern analytical techniques have identified the artificial blue pigment as calcium copper silicate and similar in composition to the rare natural mineral cuprorivaite ($CaCuSi_4O_{10}$). Modern methods of preparing the synthetic blue pigment involve fusing together a mixture of desert sand, crushed limestone, hydrated sodium carbonate (natron) and fragments of copper at temperatures of ~900°C; these materials and procedures would have been available to the ancient Egyptians.

Natural lapis lazuli pigments were traditionally known as ultramarines meaning 'from beyond the seas'; one of the best sources of natural lapis lazuli is Afghanistan, both today and thousands of years ago. Lapis lazuli is a mixture of minerals only one of which, lazurite a sulphur-containing sodium silicate, is blue. The best ultramarine blue pigment was highly prized by Renaissance artists and required a complex purification procedure to separate lazurite from other mineral components, primarily white calcite and iron pyrites, a brass-yellow mineral often called fool's gold. A cheaper alternative to the very expensive blue lapis lazuli pigment was derived from azurite [$2CuCO_3 \cdot Cu(OH)_2$] which was often found associated with copper ore deposits. Some medieval artists used two superimposed blues: azurite pigments for an under-paint and then lapis lazuli/lazurite for the top layers. Over time however, the azurite blue could attain a green hue as the pigment weathered to malachite [$CuCO_3 \cdot Cu(OH)_2$], the traditional source of green pigments.

Cinnabar [mercury(II) sulphide, HgS] was, and still is, predominantly mined as a source of mercury. Moreover, this mineral has also been valued for millennia for its red colour, being the raw material for the pigment vermilion. Vermilion was originally derived from the powdered mineral and by the Middle Ages the synthesis of the compound by heating a mixture of mercury and sulphur was well known. The synthetic pigment gave a bright vibrant red colour and was often regarded as superior to the powdered cinnabar which could vary in colour due to impurities. However, the use of mercury vapour in the synthesis made the process particularly hazardous. Vermilion was largely replaced by a new (albeit also toxic) synthetic pigment cadmium red [cadmium sulphoselenide, Cd(S,Se)], when this became commercially available in the early 20th century.

The two arsenic sulphide pigments, orpiment (As_2S_3) and realgar (As_4S_4), became largely obsolete with the development of cadmium yellow (CdS) and cadmium red [Cd(S,Se), a co-precipitate of cadmium sulphide and cadmium selenide]. Orpiment had been particularly valued for its bright golden-yellow colour in many of the ancient Eastern civilisations, despite the known toxicity associated with arsenic and its compounds. Synthetic arsenic pigments were manufactured in the 19th century although the presence of the precursors (arsenic oxides and sulphur) and sublimed products discovered in a 16th century smelter together with other historical records suggests the practice of trade in artificial orpiment preparations is centuries old.

4.3.2.3   Development of Synthetic Pigments

Investigations of the rare mineral crocoite, now known to be $PbCrO_4$, led to the discovery of the element chromium in 1798 and then in the early 19th century to the preparation of synthetic chrome yellow and later the green pigments: chrome green (chromium oxide, Cr_2O_3) and viridian green (hydrated chromium oxide). Note that chrome yellow is either lead chromate or a compound with a mixed lead chromate-sulphate composition ($PbCrO_4 \cdot PbSO_4$).

With knowledge regarding preparation of Egyptian blue having been lost over the centuries, ultramarine remained the best blue pigment available until the 18th and 19th centuries when the compounds Prussian blue and cobalt blue were created. Prussian blue, iron(III) hexacyanoferrate(II), $Fe^{III}_4[Fe^{II}(CN)_6]_3$, was first produced in 1706 and the process was kept a secret until a recipe for its preparation in 1724 revealed that it was made from heating (calcining) dried beef blood and potash (potassium carbonate). The blue pigment was adopted shortly after its invention by artists and by the early 19th century recipes for preparing Prussian blue had as their starting materials potassium hexacyanoferrate(II), $K_4[Fe(CN)_6]$. Mixing $K_4[Fe(CN)_6]$ with an iron(III) salt in aqueous solution immediately generates an inky blue precipitate of $Fe^{III}_4[Fe^{II}(CN)_6]_3$. An alternative synthesis of Prussian blue involves adding iron(II) salts to $K_3[Fe(CN)_6]$ and this is the basis of cyanotype photography which was developed in the 1840s (see Section 6.4.1).

Impure forms of cobalt ores such as cobaltite (CoAsS) and skutterudite ($CoAs_{2-3}$) had been used for centuries to impart a blue colour (CoO) to glass and porcelain with the cobalt(II) oxide forming in the high temperatures of the manufacturing processes. Experiments in 1802 with powdered cobalt glass (potassium cobalt silicate, known as smalt) and alumina resulted in the compound cobalt blue [cobalt(II) aluminate, $CoAl_2O_4$].

The original (alchemical) recipe for making Prussian blue had its origins in experiments associated with making a lake pigment. Serendipitous discoveries such as this run through many areas of science and the development of mauve (see Section 4.4), monastral blue (below) and an early synthesis of indigo (see Section 4.3.3.1) also began with chance observations from experiments that did not proceed quite as planned.[4] Insight into the puzzle as to how Prussian blue was formed in the alchemical process in 1706 came over a hundred years later. By that time cyanic acid, HCN, had been synthesised and it was concluded that calcination of beef blood in 1706 had produced HCN which had then reacted with potash to generate KCN. Subsequent reaction between KCN and iron salts, which had also been formed in the calcination process, led to Prussian blue.

An unexpected blue by-product was obtained from the reaction between ammonia and phthalic anhydride to produce phthalimide (Figure 4.13a). Investigations into the blue impurity suggested that this substance had formed because the reactants had come into contact with the iron component of the reaction container. The blue impurity was later identified as iron(II) phthalocyanine, the structure of which linked four units of phthalonitrile, $[C_6H_4(CN)_2]$, in a 16-membered ring of alternating carbon and nitrogen atoms around a central Fe^{2+} ion. Subsequent experiments showed that other metal ions could take the place of Fe^{2+} in the chemical reaction and monastral blue (or phthalo blue) is the corresponding copper(II) phthalocyanine (Figure 4.13b). Chlorinated copper(II) phthalocyanine is a green pigment in which most of the hydrogen atoms of the blue compound have been replaced by chlorine. A sulphonation reaction of copper(II) phthalocyanine converts the pigment into a water-soluble dye which is commercially used in printer inks.

4.3.2.4   Blacks and Whites

Most black pigments are carbon-based such as ground graphite or made from charring organic materials such as wood and bones and thus have changed very little over the years. Soot collected from oil lamps and fireplaces was a popular source of black in the Renaissance period, with the more recent equivalent being obtained from partly combusted natural gas. The iron ore magnetite (Fe_3O_4) is a historic black pigment and the modern counterpart is manganese ferrite ($MnFe_2O_4$) prepared by

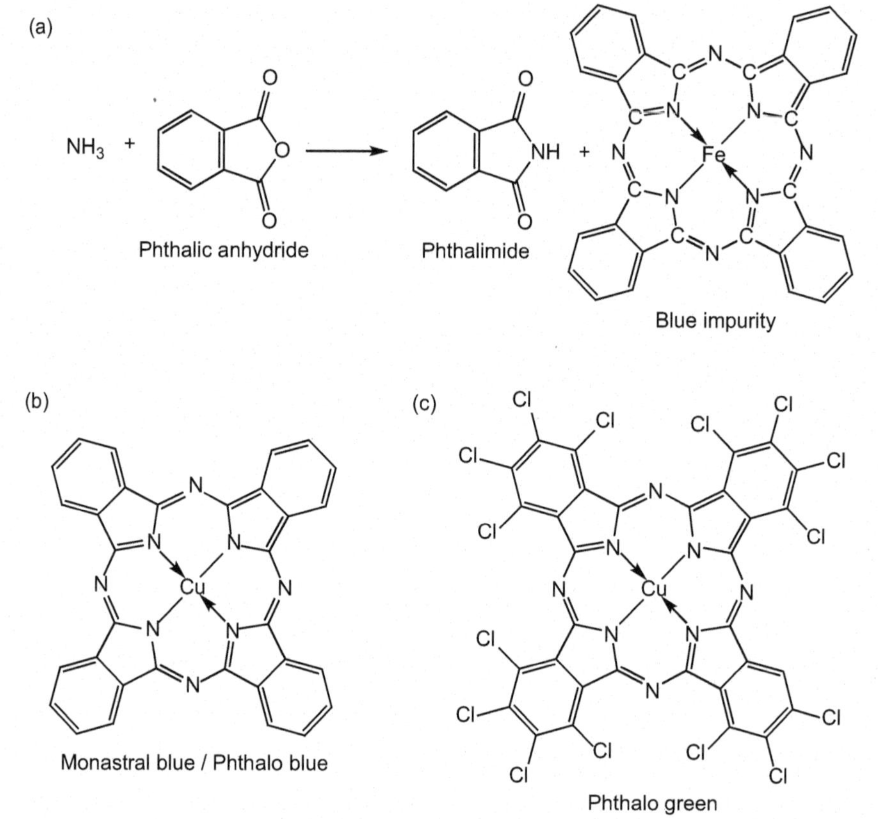

(a)

Phthalic anhydride   Phthalimide

Blue impurity

(b)

Monastral blue / Phthalo blue

(c)

Phthalo green

FIGURE 4.13 Phthalocyanine compounds. (a) Synthesis of phthalimide, a compound required for the manufacture of dyes, and the blue by-product. The 16-membered ring system of the phthalocyanines is similar to that found in porphyrin derivatives such as chlorophyll and haem. Carbon atoms in the 16-membered ring are shown for clarity in iron(II) phthalocyanine and omitted from the structures shown in (b) and (c). The metal ion (Fe^{2+} or Cu^{2+}) at the centre of the phthalocyanines is held by two covalent and two coordinate bonds to the four nitrogen atoms in the five-membered rings. (b) Blue copper(II) phthalocyanine pigment. (c) Phthalo green, the chlorinated copper(II) phthalocyanine derivative. The compound is a mixture of several different substitution products and the chemical formula ranges from $C_{32}H_3Cl_{13}CuN_8$ to $C_{32}HCl_{15}CuN_8$.

heating mixtures of manganese and iron oxides at high temperatures. Chalk, limestone and shells, the traditional sources of white pigments, have largely been superseded by artificial compounds. Zinc white (ZnO) was introduced in the 18th century, titanium white (TiO_2) in the early 20th century whilst the original artificial white pigment, flake white ($PbCO_3$), has largely been rendered obsolete due to its toxicity and the availability of suitable alternatives. Titanium white is a key ingredient in many white wall paints and some DIY outlets will add coloured pigments to the base material to give customers the precise shade and colour desired.

4.3.2.5 Organic Pigments

Apart from the phthalocyanines, the pigments discussed above are inorganic compounds. Organic azo compounds developed in the late 19th and early 20th centuries are predominantly used as soluble dye molecules (see Section 4.3.3) and contain one or more –N=N– groups. However, some azo compounds are insoluble and can be used as pigments if the molecule lacks any solubilising sulphonic acid (–SO$_3$H) groups or if it is converted into a lake pigment through reaction with a metal salt. Azo pigments dominate the yellow, orange and red parts of the colour spectrum and are largely used as paints, as printing inks and to colour plastics. Different azo compounds can be mixed to give a specific shade to a particular colour. An imitation vermilion pigment, for example, is composed of a mixture of a red and yellow azo compound (Figure 4.14a and b) together with white barium sulphate. A range of different substituents (e.g. CH$_3$, Cl, NO$_2$ etc.) can be incorporated into the structures of the starting materials (see Figure 4.21) to subtly change the shade or properties of the molecule. Quinacridones are red compounds composed of five interlinked rings with two pairs of oxygen and nitrogen atoms. Substitution of hydrogen atoms of the aromatic rings with other groups, as in the azo compounds, changes the colour and quinacridone magenta pigment has methyl groups on the end rings (Figure 4.14c).

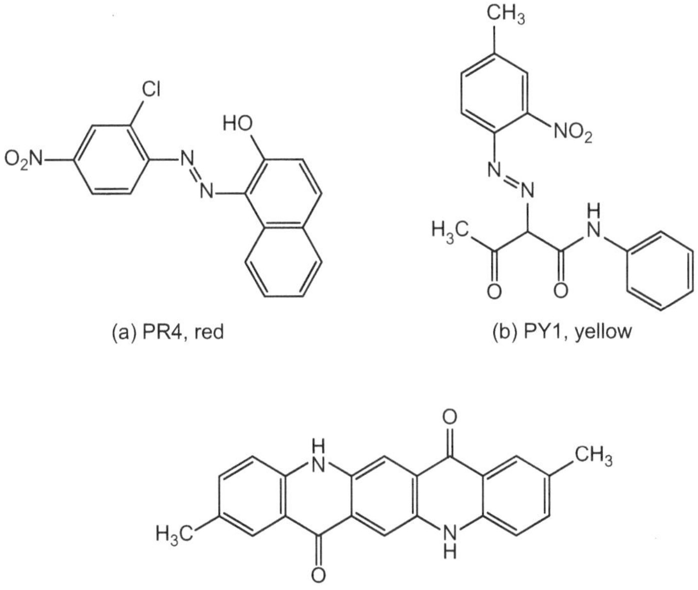

(a) PR4, red

(b) PY1, yellow

(c) PR122, magenta

FIGURE 4.14 Azo and quinacridone compounds. PR4 is a β-naphthyl azo compound, PY1 is an azo acetoacetic arylide and PO5 (structure not shown) is orange and a similar molecule to PR4 but with a –NO$_2$ group in place of the chlorine substituent. The parent acridone structure with no methyl groups is a more violet shade of magenta. Each coloured compound is given a colour index (C.I.) number: PR4, C.I. 12085; PY1, C.I. 11680; PO5, C.I. 12075 and PR122, C.I. 73915.

4.3.3   DYES: NATURAL AND SYNTHETIC

4.3.3.1   Indigo and Related Compounds

One of the oldest plant-based blue colourants is indigo. Historically, indigo has been used both as a pigment in paintings and as a dye to colour fabrics and is most familiar now as the dye used to colour blue jeans. Indigo is produced from the processing of leaves of the *Indigofera* genus of plants, such as woad, during which a glycoside precursor molecule is converted to the blue compound (Figure 4.15). Indigo is insoluble in water, and thus can act as a pigment, but reduction to leucoindigo or 'white indigo' gives a water-soluble compound and it is this form which is applied to fabrics. After the fabric has been suitably impregnated with the water-soluble compound it is exposed to atmospheric oxygen which causes the leucoindigo to reoxidise to insoluble indigo and regeneration of the blue colour. Originally the reduction of indigo was done with urine, containing ammonia as the reductant, or by using certain species of anaerobic bacteria but modern methods use sodium dithionite ($Na_2S_2O_4$) or thiourea dioxide [$H_2N-C(=NH)-SO_2H$]. Indigo therefore becomes physically deposited within the fibres and becomes trapped because of the molecule's insolubility. Indigo however is a small molecule and, despite its insolubility, this dye is susceptible to

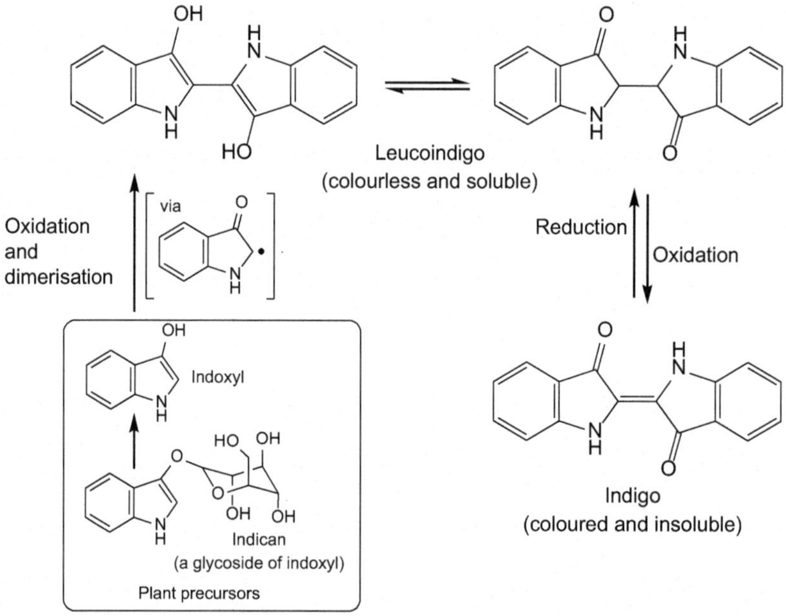

FIGURE 4.15   Indigo dye. Air oxidation of the indoxyl plant precursor generates a radical, a species with an unpaired electron, which dimerises to form leucoindigo and then oxidation of this compound gives indigo. During the fabric dying process the indigo is first reduced to leucoindigo in an alkaline deoxygenated solution and then the fabric is immersed in this solution. The fabric is left soaking for a period of time and upon removal it changes to a blue colour as it oxidises.

being removed during washing, and dyed fabrics can acquire a faded appearance with time.

Indigofera plants are known to have been grown and used by ancient civilisations such as the Ancient Egyptians and Babylonians to produce the blue colour and it was an important trading commodity for centuries. European demand for indigo in the 19th century was largely met through large-scale cultivation and processing of *Indigofera tinctoria* plants in India but production of natural indigo dropped significantly in the early 20th century after synthetic routes were developed.[5] Synthetic chemical methods are able to not only give the identical structure to the natural molecule but can be adapted to produce a variety of related compounds with different properties and/or colours (Figure 4.16). Reaction of indigo with sulphuric acid incorporates sulphonic acid groups (SO_3H) on the aromatic rings and the sodium salt derivative, indigo carmine (E132), is a water-soluble blue/green colourant used in the food, pharmaceutical and cosmetic industries. Removal of the internal N–H------O=C hydrogen bonds in indigo by replacing the NH groups by sulphur significantly changes the chemical characteristics of the compound and thioindigo is a pink dye for polyester fabrics.

Purple has a long association with royalty as it was an extremely expensive dye to produce in antiquity, being obtained from the glands of *Murex* species of Mediterranean sea snails in a long, difficult and laborious process. The major component of the pigment, Tyrian purple (see Figure 4.16), is now known to be 6,6′-dibromoindigo but small amounts of other natural pigments present in the molluscan extract can vary the colour of dyed cloth between crimson and deep purple. The chance discovery of a purple dye (mauve) by the young chemist William Henry Perkin in 1856 from cheap and readily available starting materials led the way to the synthetic dye industry (see Section 4.4).

4.3.3.2 Chlorophyll and Chlorophyllins

The most important compound in plants is chlorophyll (Figure 4.17) as it not only gives leaves their green colour but is also responsible for initiating the photosynthesis

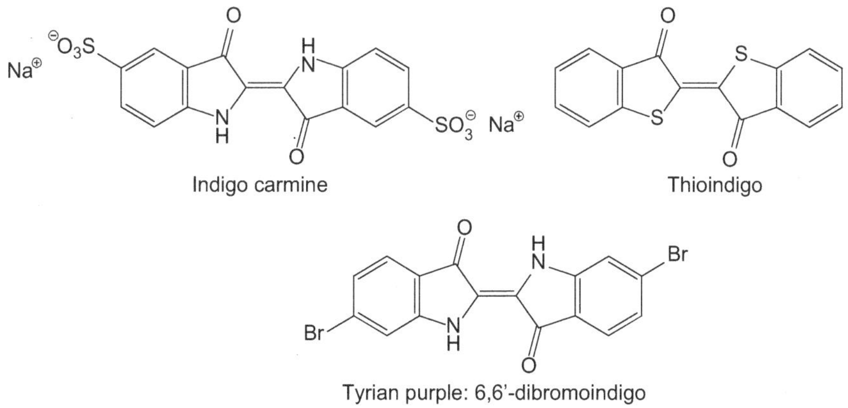

Indigo carmine

Thioindigo

Tyrian purple: 6,6'-dibromoindigo

FIGURE 4.16 Indigo derivatives: indigo carmine, thioindigo and Tyrian purple.

process which produces glucose and oxygen from carbon dioxide and water. The chlorophyll molecule strongly absorbs the red and blue wavelengths of sunlight and thus the light it reflects and transmits to the eye is green. Absorption of the light causes an electron from the porphyrin ring in chlorophyll to be excited from a lower energy state to a higher energy state. The excited electron is then passed along an electron transport chain which culminates in generating the molecules ATP and NADPH which feed into the Calvin cycle to produce glucose.

Whilst the most important role of chlorophyll is in photosynthesis, this compound can be extracted from plant sources such as grass and nettles and used as a green colouring agent (E140) in foods, toothpaste, soaps and cosmetics. The total synthesis of chlorophyll in the middle of the 20th century by R. B. Woodward and co-workers was an outstanding achievement in organic synthesis and involved many elegant reaction sequences. The synthesis is long and complex and not economically viable and thus the chlorophyll used as a colourant is the material extracted from natural product sources. However, chlorophyll pigments are not very stable and the long hydrocarbon tail ($C_{20}H_{39}$) can be easily cleaved both enzymically and non-enzymically. Hydrolysis of chlorophyll extracts with sodium hydroxide solution and then addition of copper salts results in semi-synthetic chlorophyllins (E141, Figure 4.17). These compounds retain the core macrocyclic ring system responsible for the green colour but are water soluble and are used as colourants in the drinks industry where the hydrophobicity of chlorophylls is a source of difficulty.

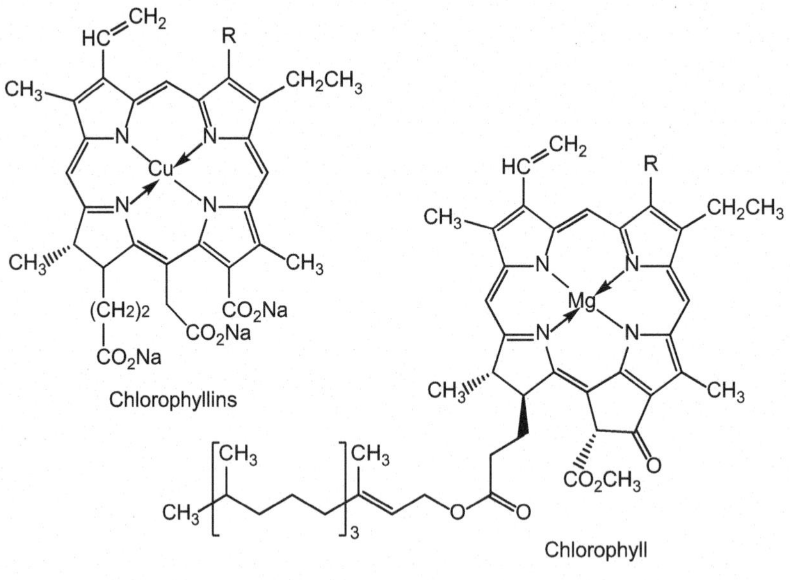

FIGURE 4.17 Green colouring agents. Chlorophyll is a mixture of two compounds, chlorophyll-*a* (*R* = CH_3) and chlorophyll-*b* (*R* = CHO) in the ratio 3 to 1 respectively. These *R* groups are retained in the water-soluble green chlorophyllin dyes. The metal ions, Cu^{2+} and Mg^{2+}, are incorporated at the central positions of the chlorophyllins and chlorophyll respectively, and are held by two covalent and two coordinate bonds to the nitrogen atoms in the five-membered rings.

4.3.3.3   Alizarin, Carminic Acid and Carmine

Many traditional natural red dyes are based on an anthraquinone structure, i.e. two aromatic rings joined by two carbonyl groups to give a third ring. Alizarin (Figure 4.18a) was obtained from the roots of the common madder plant *Rubia tinctorum* but dyed cloths quickly faded and the compound only sticks fast to fabrics which have first been impregnated with a mordant. The mordant is commonly a metal salt which binds to the fibres of the fabric and then reacts with the alizarin dye to create

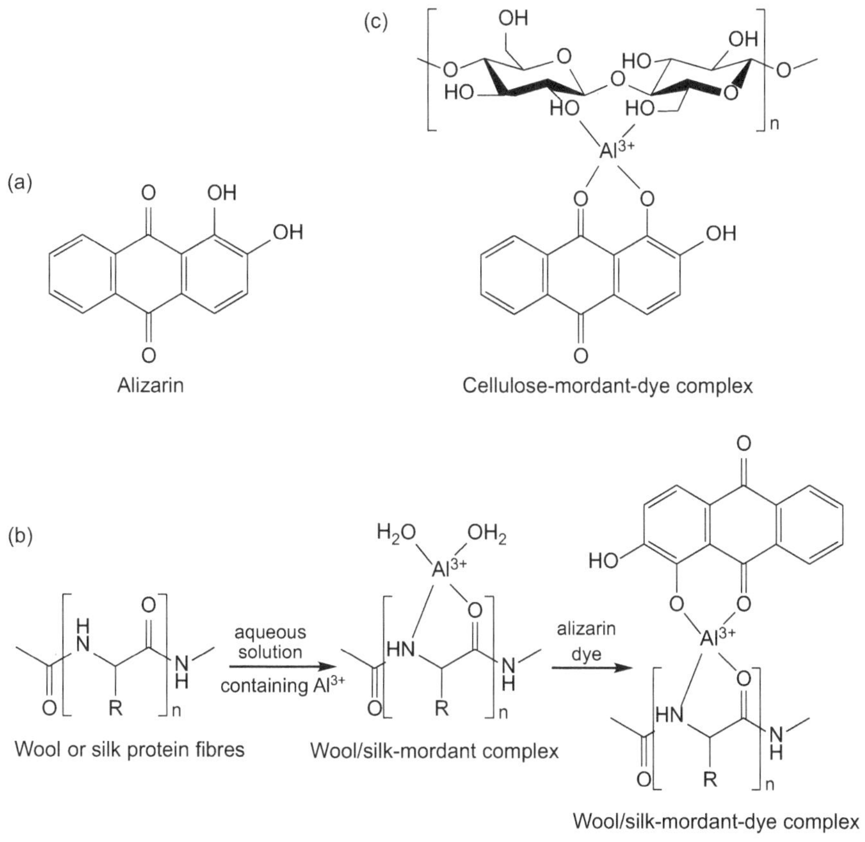

FIGURE 4.18   (a) Alizarin and fibre-mordant-dye bonding via a chelate ring in (b) protein structures in wool or silk and (c) cellulose (cotton). The remaining two coordinated ligands to Al^{3+} are H_2O or OH^-. Mordanting under alkaline conditions results in more intense colours; modern procedures use sodium carbonate solutions (given as soda ash in old recipes) whilst stale urine was used in ancient times. Protein fibres are more receptive towards mordants than those in cellulose and the dyeing of cellulose textiles sometimes requires a two-step mordanting procedure, first with tannins and then with alum. Tannins are naturally occurring soluble polyphenolic compounds of high molecular weight (around 500–3,000) and are better than alum in forming bonds to cellulose. In the first mordanting step the phenolic hydroxyl groups of the tannin (often obtained from oak galls) form bonds with the hydroxyl groups of the cellulose and then this tannin-fibre complex crosslinks with the alum in the second step.

insoluble compounds of the dye bonded to the fibre (Figure 4.18b). Mordanting is often essential for the dyeing of cellulose fabrics such as cotton and linen but also helps with improving the fastness of all dyed fabrics. Potassium alum [potassium aluminium sulphate, $KAl(SO_4)_2 \cdot 12H_2O$] was the most important mordant in ancient times, being obtained from mineral sources such as alunite [$KAl(SO_4)_2 \cdot 2Al(OH)_3$]. Modern preparations of alum involve adding potassium sulphate to a concentrated solution of aluminium sulphate, a compound which is obtained from the treatment of the aluminium ore bauxite (hydrated Al_2O_3) with sulphuric acid. Aluminium ions, in common with first row transition metal ions, form octahedral complexes with electron-donating molecules (ligands) such as water. During the mordanting step electron-donating groups on the fibres substitute for two of the ligands of the hexaaqua aluminium ion, $[Al(H_2O)_6]^{3+}$, thus firmly attaching the aluminium ions to the fabric. A similar process occurs during the subsequent reaction with the dye molecules resulting in a stable link between fibre and coloured compound.

Alizarin was the first natural product to be synthesised in the laboratory and the route started from anthracene, a waste material from the coal tar industry (Figure 4.19). The discovery of a commercially viable synthetic route in the late 1860s was a significant achievement as the molecular structure of alizarin was only

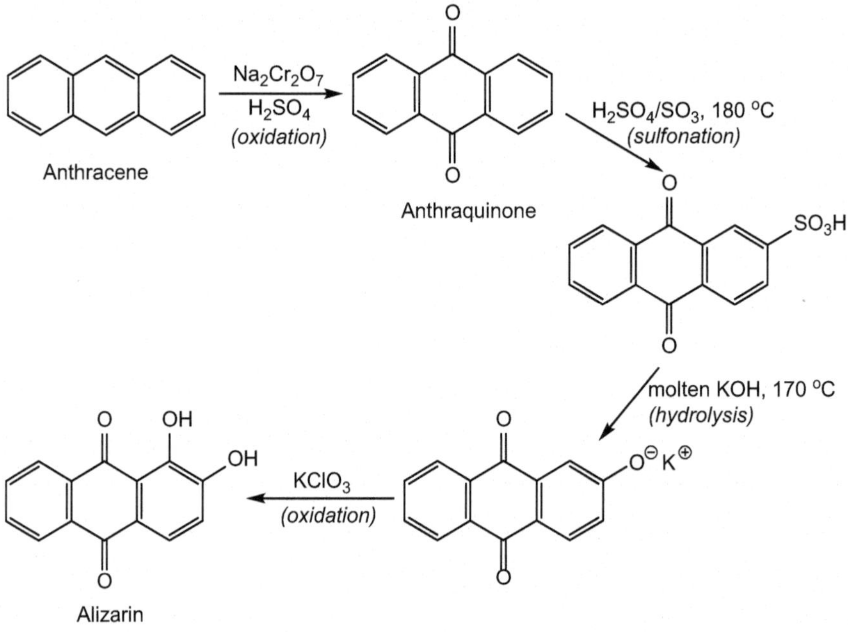

FIGURE 4.19 An early commercial synthesis of alizarin (Perkin). Two alternative synthetic routes were developed, one by Carl Graebe and Carl Liebermann from the German BASF company and the second by William Henry Perkin in London. The BASF patent was filed on 25 June 1869 in London, one day before Perkin's on 26 June 1869. A potential patent dispute was avoided by allocating Perkin the British market whilst those in mainland Europe and the United States were given to BASF.

elucidated in full several years later. Once the synthetic compound was commercially manufactured it soon dominated the alizarin dye market and cultivation of the madder plants declined with significant economic repercussions for the farmers, a situation which was echoed subsequently with indigo production.

Carminic acid (Figure 4.20) is a mordant dye extracted from female cochineal insects (*Dactylopius coccus*) native to South America which results in bright crimson fabrics. The dye was introduced into Europe in the 16th century and gave a superior colour to the red Mediterranean dyes from the scale insect *Kermes vermilio*. Carminic acid is a pH-dependent molecule as there are many ionisable OH groups and this can generate variations in the colour. An orange colour can be obtained by adding citric acid to the dye solution whilst purple shades result from the addition of sodium carbonate. Some soluble dyes are reacted with metal salts and converted into insoluble compounds which are sometimes referred to as lake pigments. The

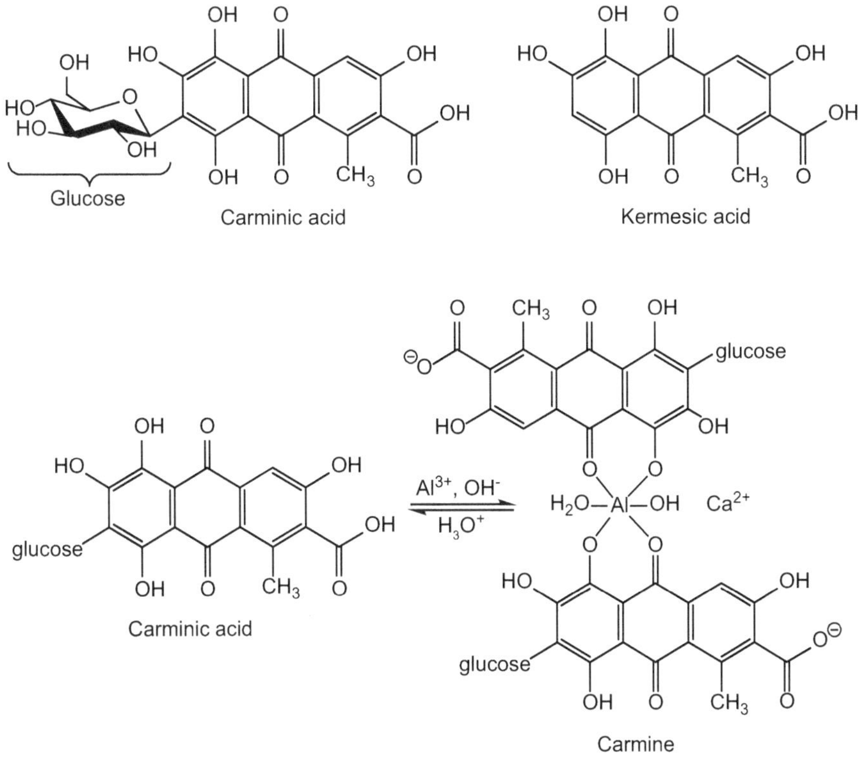

FIGURE 4.20 Anthraquinone insect dyes: carminic acid, kermesic acid and carmine. In the insect the carminic acid is thought to be a form of defence against predators. Kermes insects produce kermesic acid which is the aglycone of carminic acid. Carmine is a synthetic lake pigment consisting of a metal ion complexed with two carminic acid molecules at their central carbonyl and adjacent hydroxyl groups. The carminic acid molecules act as bidentate ligands, forming two coordinate bonds with the metal. Carmine is less pH dependent than carminic acid and gives a consistent red colour between pH 4 and 10.

lake pigment carmine is the precipitated aluminium complex of carminic acid and is obtained by adding aluminium sulphate to filtered cochineal extracts under basic conditions. In acidic conditions carmine can be converted back into carminic acid. Carminic acid and the insoluble lake pigment carmine are currently most commonly used commercially as food colourants and in cosmetics.

4.3.3.4   Azo Dyes

Azo compounds are synthetic compounds with a backbone comprising of at least one –N=N– bond linking aromatic ring systems (Figure 4.21). It is this extended π system and the auxochrome substituents (e.g. $-CO_2H$, $-NH_2$, $-SO_3H$, $-OH$) on the aromatic rings which gives these molecules a strong absorption in the visible region. Azo dyes give predominantly yellow/orange/red colours with relatively few greens and blues. Most azo dyes are synthesised by diazotisation of an aromatic primary amine followed by coupling with an electron-rich aryl nucleophile such as a phenol (Figure 4.21a). Azo compounds used for fabric dyes generally have one or more sulphonic acid groups to increase water solubility and to provide bonding between the dye and the surface of the fabric (Figure 4.21b and c).

Sulphonic acid groups are strong acids (and so exist as sulphonate ions at pH 7) and the anionic dyes such as acid orange and acid red form ionic bonds to protonated amino groups (the cationic sites) in animal protein fibres (wool and silk):

$$Dye - SO_3^-Na^+ + Wool - NH_3^+Cl^- \rightarrow Dye - SO_3^- {}^+NH_3 - Wool(+NaCl)$$

Polyacrylonitrile fibres carry a negative (anionic) charge in their backbone making the ionic character of the textile-dye interaction the reverse of that for the acid dyes.[6] An example of a cationic dye for acrylic fibres is basic red 18 (Figure 4.21d) which has a quaternary ammonium centre $[-^+N(CH_3)_3]$. Cationic dyes can also be used to dye wool and silk as these are protein fibres and contain carboxylate $(-CO_2^-)$ groups.

Direct dyeing of cotton and linens is inherently less successful than with wool and silk as only weak intermolecular interactions can form between the cellulose hydroxyl groups and the azo dyes compared to the stronger ionic bonding possible with protein fibres. Moreover, dyeing is a water-based process and it is difficult to ensure that azo dye molecules which adhere to cellulose fibres through hydrogen bonding remains attached to the fabric and do not wash out. Azo dyeing of cellulose textiles largely uses reactive dyes which undergo a reaction with the fibres to form a covalent bond (Figure 4.22). By forming a permanent attachment to the textile fibres these dyes give excellent colour fastness to wash and were an immediate commercial success when introduced in the 1950s.

The synthesis of azo dyes takes place in two steps, diazotisation and coupling, and when this procedure is carried out in the presence of cellulose fibres to generate an insoluble coloured compound, this product can become trapped inside the fibres in a manner analogous to dyeing of indigo. For example, the azo product Sudan 1 (C.I. 12055) from the synthesis in Figure 4.21a and Pigment red 22 (C.I. 12315) (Figure 4.21e) can be used in this manner.

Azo dyes are used widely as food colourants in addition to dying textiles and there are regulations governing their use. Some types of azo dyes (fewer than 5%)

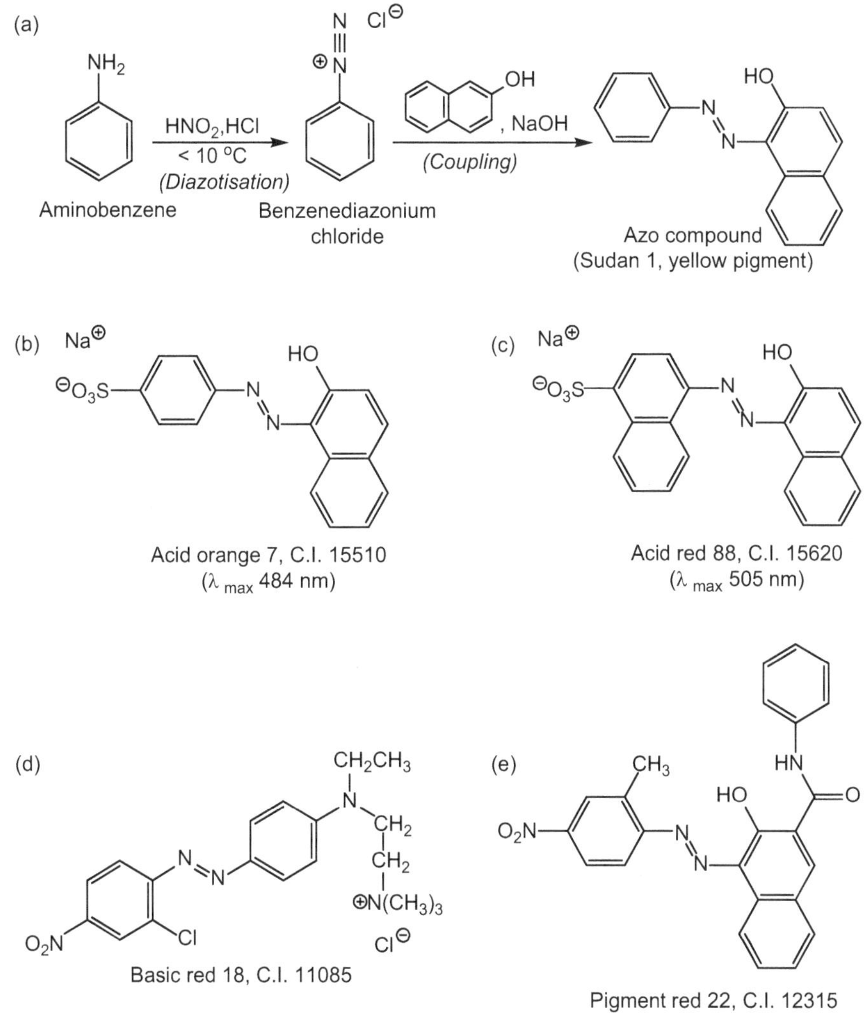

FIGURE 4.21 Azo dyes. (a) The basic two-step synthetic method of preparing azo compounds is diazotisation followed by a coupling reaction. (b) Acid orange 7 is synthesised by coupling β-naphthol (a homologue of phenol) and the diazonium derivative of sulphanilic acid (HO₃S–C₆H₄–NH₂). (c) Acid red 88 is obtained from the starting materials naphthionic acid (HO₃S–C₁₀H₆–NH₂) and β-naphthol. (d) Basic red 18 is a cationic dye. (e) Pigment red 22 a water-insoluble dye is formed by coupling the two reactants on the fibre and is therefore referred to as an azoic dye. Azo dyes are the largest category of dyes (more than 2,000 compounds) and account for 60%–70% of all dyes used in industry. There are a range of basic structural types in addition to those shown above and in Figure 4.14.

have been withdrawn from use because they can be absorbed, if in direct and prolonged contact with the skin or mouth, and then be metabolised to produce harmful or carcinogenic aromatic amines such as benzidine, $H_2N–C_6H_4–C_6H_4–NH_2$. Waste water discharged from textile industries contains unreacted dyes and other pollutants

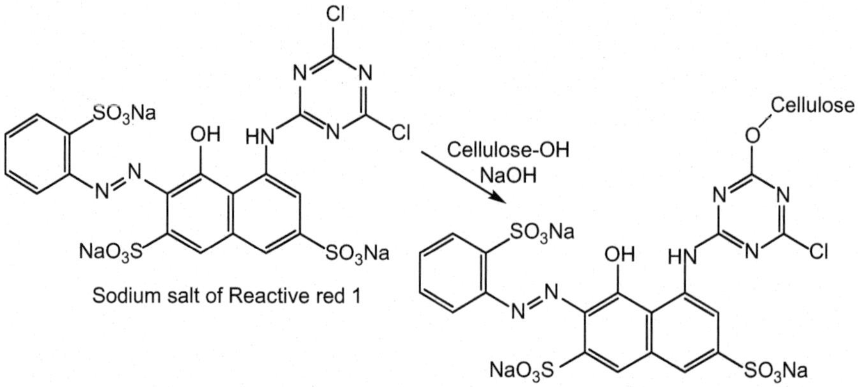

FIGURE 4.22 Dye-cellulose fixation reaction with Reactive red 1 (C.I. 18158). Reactive dyes get fixed into the fabric by nucleophilic substitution of chlorine atoms in the dichlorotriazine (the 'reactive group') part of the molecule with hydroxyl groups of the cellulose under alkaline conditions. Reactive dyes also react with water causing hydrolysis and become inactive. This hydrolysis reaction can be reduced by, for example, keeping the temperature of the dyeing solution <40°C or by applying the pre-dissolved dye to the cotton prior to the addition of alkali. Wool and silk can also be dyed using reactive dyes but require milder alkaline conditions to protect the more delicate fibres.

and various methods are employed to remove these substances but they are not completely effective. Better treatment protocols are needed to minimise the environmental impact and health hazards of textile dye waste and a variety of research programmes are continuing to explore new procedures and technologies.

4.3.3.5 Gold Nanoparticles

Lateral flow tests (LFTs) for the rapid detection of coronavirus particles became part of everyday life for millions during the Covid pandemic. These tests were in effect sandwich immunoassay kits which gave a qualitative readout within 30 minutes (Figure 4.23). Capillary action caused the sample solution to move along the membrane strip of the LFT where it would first interact with gold nanoparticles coated with antibodies that recognised spike proteins (antigens) on the surface of virus particles. The sample then continued onto the T and C lines which were printed with 'capture' receptors. Embedded at the T line were antibodies that would bind to a different part of the virus particle to that recognised by the nanoparticle-antibody complex and at the C line were antibodies that would bind to antibodies on the nanoparticles. If virus particles were present, they would accumulate at the T line through the binding of the antigen-antibody-nanoparticle complex to the capture receptor, and unbound nanoparticle-antibodies would be detected at the C line. The red/pink colour at the test (T) and control (C) lines was generated by the surface plasmon resonance (SPR) effect of the gold nanoparticles.

Free electrons can generally move freely through the crystalline structure of a metal but in metal nanoparticles the surface/volume ratio becomes high and a large proportion of atoms are located on the surface. Movement of the free electrons then

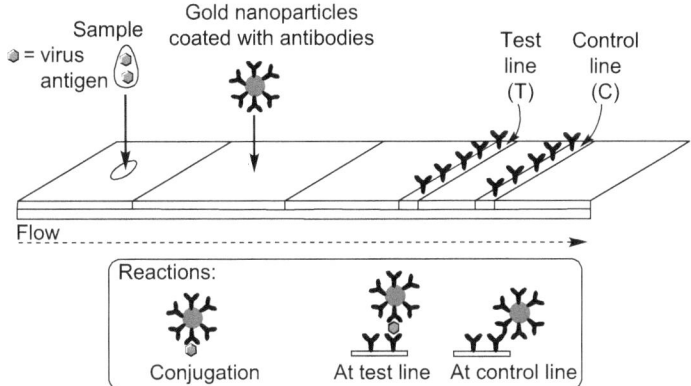

FIGURE 4.23   Schematic of Covid lateral flow test.

becomes confined to a short length and when metal nanoparticles are irradiated by photons, a portion of the light energy interacts with the free electrons causing a concerted oscillation of electric charge. When the frequency of the incoming light is resonant with the coherent electron motion a strong absorption occurs which is called SPR. For nanoparticles of copper, silver and gold the SPR appears in the visible part of the electromagnetic spectrum. The SPR of small gold nanoparticles (<30 nm) causes absorption of blue-green light whilst red light is reflected giving a red colour. As the particle size increases the SPR absorption shifts to longer, redder, wavelengths changing the observed colour from red to blue and purple.

Gold nanoparticles can be prepared by breaking down bulk gold or building up from the atomic level by reduction of gold precursor ions in solution. In the most common synthetic route which was first described in 1951, chloroauric acid, $HAuCl_4$, is dissolved and boiled in water and then an aqueous solution of trisodium citrate is added with vigorous stirring. After a few minutes a wine-red colloidal suspension of gold is observed with a particle size of around 20 nm. The citrate ions also act as a capping agent to prevent aggregation of nanoparticles and formation of a metallic precipitate. The citrate/Au^{3+} ratio controls the particle size of the gold nanoparticles; a higher relative concentration of the citrate decreases the gold nanoparticles size because of the stabilising effect of the reagent. Biosynthetic routes to metal nanoparticles are becoming common and various biomolecules in plants and bacteria and fungi can act as reducing and/or capping agents. The procedure usually involves incubating a solution of $HAuCl_4$ with a biological extract and this approach has the advantage over the chemical route as being more environmentally friendly.

The intrinsic properties of gold nanoparticles are essentially determined by the type of motion its electrons can have which is related to their degree of confinement and this in turn is associated with their size and shape. Generally bulk gold is regarded as chemically inert; however, gold nanoparticles smaller than 10 nm show catalytic activity for many reactions of industrial and environmental importance such as the selective oxidation of alkenes and the reduction of atmospheric nitrogen oxide pollutants. The catalytic activity is a result of the high fraction of surface atoms which have a low coordination number and are sites of increased chemical reactivity.

The list of applications of gold nanoparticles is growing rapidly and includes electronics, medical treatments, biomaterials and sensors in addition to catalysis and the diagnosis of a range of diseases.

Gold nanoparticles are not a new phenomenon. The modern era of gold nanoparticles started with the work of Michael Faraday in the mid-1850s who observed that colloidal gold solutions had properties different to those of bulk gold. However, nanoparticles have a much longer history and date back to at least ancient Greek and Roman eras when they were known as ruby gold and were used to stain glass and ceramics. One of the best-known examples is the Lycurgus Cup, a fourth-century Roman drinking vessel which is displayed in the British Museum. The cup exhibits a dichroic optical effect; a jade green colour is seen when the cup is illuminated from the front but a ruby red colour when light shines through it. The presence of 50–100 nm diameter gold and silver nanoparticles, which were identified by electron microscopy, absorb light at around 520 nm thus giving the red-orange colour when the cup is viewed under transmitted light. Some of the vibrant red stained glass church windows from the medieval period are thought to owe their colour to gold nanoparticles but this is debatable and the colour could be due to a copper-based species. Knowledge of ruby gold production has been lost and rediscovered several times over the centuries but Johannes Kunckel (circa 1637–1703) is often described as the first glassmaker to produce ruby gold on a large scale. One of Johannes Kunckel's masterpieces, a ruby gold glass goblet, is in the Rijksmuseum Amsterdam.

The alchemists of the past did make gold, albeit in the form of nanoparticles, and from a reduction rather than a transmutation process. One of the main preoccupations of the medieval alchemists was to search for the philosopher's stone, the legendary substance by means of which base metals could be converted into gold. Whilst this quest did not succeed, the alchemists' work resulted in knowledge of the physical properties and chemical reactions of a huge variety of substances and laid the groundwork for modern chemistry.

4.4 PERKIN, MAUVE AND THE BEGINNING OF THE SYNTHETIC DYE INDUSTRY

During the Easter holidays in 1856 the talented 18-year-old chemistry student William Henry Perkin was attempting to synthesise the antimalarial drug quinine in his home laboratory in East London. Synthetic chemistry in the mid-19th century was in an early stage of development and no one knew that molecules had shapes and particular arrangements of atoms. One popular synthetic approach at that time was the 'additive and subtractive' method which involved comparing the (known) molecular formula of the desired product with that of a suitable starting material. The formula of the starting material could then be altered by adding or removing particular atoms through the judicious use of appropriate reagents to give the desired product. Perkin, under the direction of his mentor Professor Hofmann, proposed to synthesise quinine ($C_{20}H_{24}N_2O_2$) from N-allyltoluidine ($C_{10}H_{13}N$), a cheap and readily available compound from coal tar. The synthesis aimed to add the missing oxygen atoms from the oxidising reagent potassium dichromate ($K_2Cr_2O_7$) with the excess hydrogen atoms being removed in the form of water (Figure 4.24).

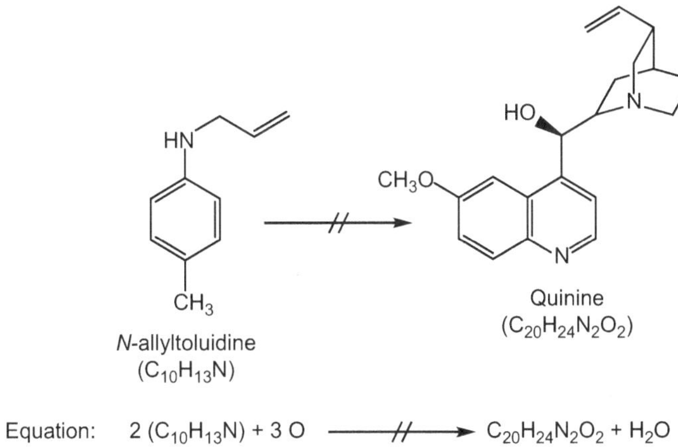

Equation: $2 (C_{10}H_{13}N) + 3 O \quad \xrightarrow{\;\;/\!/\;\;} \quad C_{20}H_{24}N_2O_2 + H_2O$

FIGURE 4.24 Proposed additive and subtractive synthesis of quinine from N-allyltoluidine. The structure of quinine was established by P. Rabe in 1908 and the first total synthesis was reported in 1945 by R. B. Woodward and W. E. Doering. Other syntheses of quinine have since been reported although none are commercially viable. Quinine is still used as an anti-malarial but other compounds such as chloroquine (a synthetic 4-aminoquinoline derivative) and artemisinin (extracted from a Chinese herb) are more widely used in treatment protocols due to fewer side effects.

Perkin did not obtain quinine from this procedure but produced a rather unprom-ising reddish powder. However, his scientific training prompted him to repeat the experiment using a different coal tar extract, one containing the simpler compound aniline, and this produced a black powder. Perkin noticed that the crude black powder partially dissolved in methylated spirits to give a purple solution. Intrigued by this unexpected finding, and aware that the textile industry was looking for new dyes, he dipped a piece of white silk into the solution and found that the colour stuck fast to the fabric. Perkin had thus accidentally discovered a purple compound which could dye silk mauve, a colour which had hitherto had only been obtainable using extracts of *Murex* species of Mediterranean sea snails. Perkin sent samples of his synthetic dye to a highly regarded dye works in Scotland for an assessment of the commercial viability of the dying procedure on different fabrics. Dyeing silk and wool was not a problem and cotton could be dyed if the fabric was pre-treated with a mordant. By the end of 1857 Perkin had obtained a patent for his synthetic compound, left his studies at the Royal College of Chemistry in London, solved many problems associated with scaling up his laboratory procedure to an industrial scale and set up a factory with his father and brother to manufacture the new dye. Perkin reported that he used at least two methods for the production of the commercial mauve dye that afforded '*two different products, namely a blue shade of mauve prepared from aniline containing but little toluidine, and a red shade prepared from an aniline containing large quan-tities of toluidine*'. Chemical formula data obtained by Perkin indicated the presence of four nitrogen atoms in the mauve dye and either 26 or 27 carbon atoms which was consistent with the presence of two similar compounds in varying amounts. Modern analysis of Perkin mauve samples from museum collections has shown that the dye was a complex mixture of several related compounds with the major components being mauveine A and mauveine B (Figure 4.25).

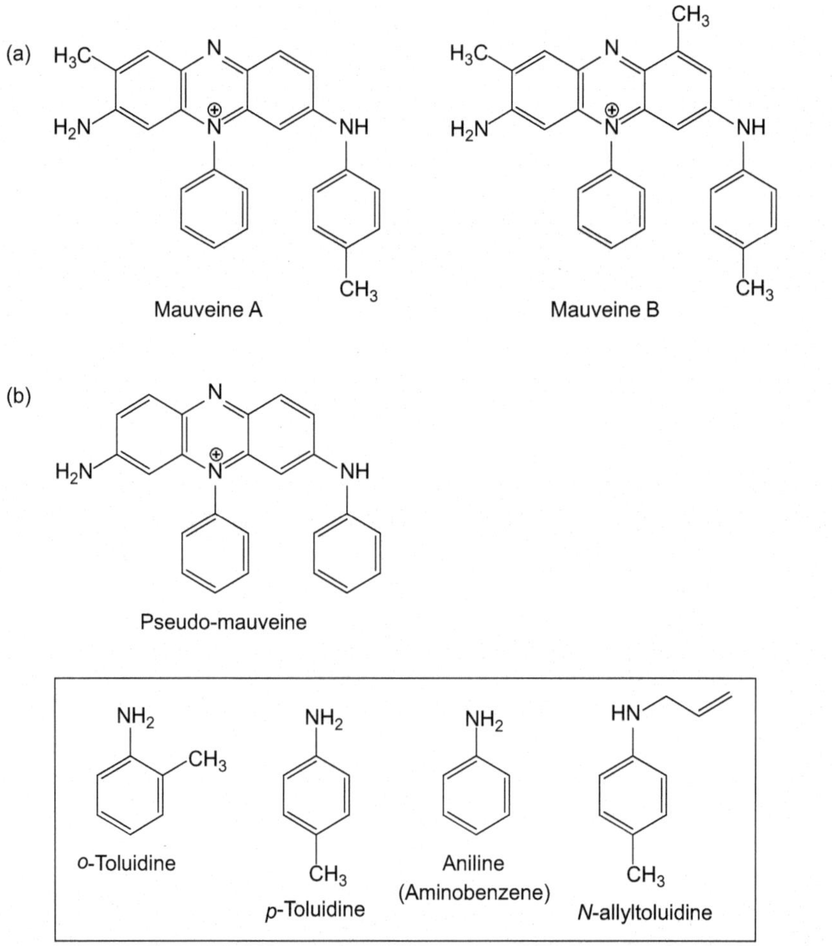

Mauveine A

Mauveine B

Pseudo-mauveine

o-Toluidine

p-Toluidine

Aniline
(Aminobenzene)

N-allyltoluidine

FIGURE 4.25 (a) Mauveine A and mauveine B are the di- and trimethyl derivatives respectively of a non-methylated core compound pseudo-mauveine shown in (b). This latter compound was described by Perkin as a second, more soluble, colouring matter in the mauve dye with the formula $C_{24}H_{20}N_4$ [correct formula $C_{24}H_{19}N_4^+X^-$, X^- = acetate $(CH_3CO_2^-)$] and in modern synthetic procedures it can be prepared from pure aniline, $C_6H_5NH_2$. Note that all the coloured compounds were obtained as salts with the counterions being either sulphate (SO_4^{2-}), chloride (Cl⁻) or acetate $(CH_3CO_2^-)$. *Inset.* In Perkin's original experiments aimed at preparing quinine he started with N-allyltoluidine from coal tar. However, it was fortuitous that this starting material was impure and it contained other aromatic compounds such as aniline and two isomers of toluidine which were the key components for the preparation of mauve. Each of the mauveine compounds in the mauve dye contains four units of one or more of the aniline-based starting materials.

Mauve-coloured clothes became the height of fashion for five years between 1858 and 1863 and Perkin earned a fortune. Perkin's success stimulated other dye manufacturers in the UK and Europe to further experiment with coal tar components and many new dyes were soon marketed. One molecule of aniline and two of toluidine gave a red dye [rosaniline or fuchsine ($C_{20}H_{19}N_3 \cdot HCl$), 1859, Figure 4.26], with further reactions leading to a blue dye [aniline blue, a triphenyl derivative of rosaniline (1861)] and Hofmann's violet (triethylrosaniline, 1863). Perkin himself introduced Britannia violet, a methylated derivative of rosaniline in 1874, Perkin's green (an acetylated product of Britannia violet) in the late 1860s and several compounds based on mauve. The ethyl derivative of mauve resulted in a redder colour (dahlia) whilst loss of the p-tolyl moiety gave a compound known as aniline pink or parasafranine.

One of the aims of the chemical industry in the 19th century was to serve social purposes; recall the original aim of Perkin's experiment was to artificially produce quinine which was in short supply. Experiments in the laboratory to form or decompose compounds were also helping 19th-century chemists to discover fundamental principles of organic chemistry such as the tetravalent nature of carbon and the way in which atoms were arranged in molecules. Establishing the structure of a compound in the 19th century was difficult, and even the way in which molecules are drawn and are so familiar to chemists in the 21st century was in its infancy. Natural products were different to compounds such as the dyes prepared by Perkin and other chemists in one crucial way; their structures could be established if they were also prepared synthetically in the laboratory and be shown to have identical physical and chemical properties to the natural material. Moreover, synthesising compounds for which there was already a thriving market provided entrepreneurs of the time with an attractive business opportunity. An early target was alizarin.

In 1869 Perkin and the BASF company in Germany separately obtained patents for the chemical synthesis of alizarin from the coal tar compound anthracene (see Figure 4.19). Although the precise structures of these compounds were not known with certainty at the time, it had been shown that the natural alizarin dye could be reduced to anthracene with zinc dust. Several years earlier the cyclohexa-1,3,5-triene structure for C_6H_6 (benzene) (Figure 4.27a) had been proposed and a fused tricyclic

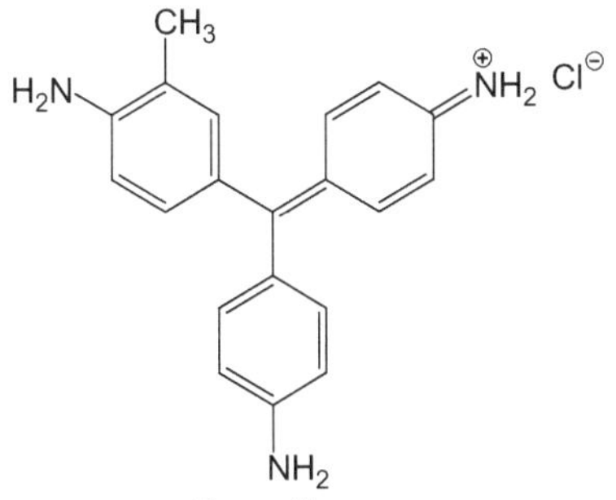

Rosaniline

FIGURE 4.26   Structure of rosaniline.

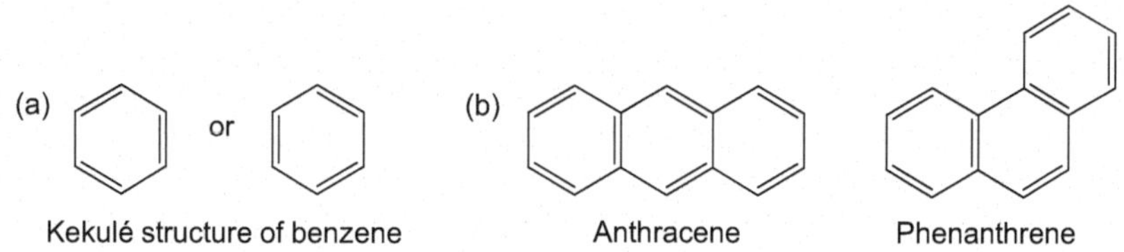

FIGURE 4.27 (a) Kekulé cyclohexa-1,3,5-triene structure of benzene with alternating single and double bonds, (b) linear (anthracene) and angular (phenanthrene) arrangements of three aromatic rings.

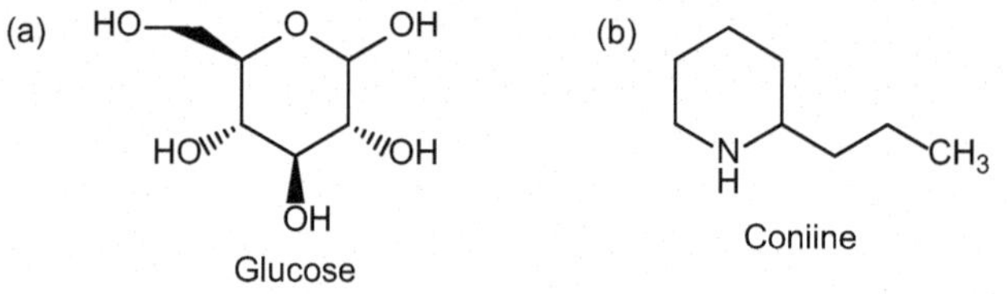

FIGURE 4.28 Early examples of synthetically produced natural products.

aromatic structure for anthracene could be visualised (Figure 4.27b). Whether a linear or angular arrangement of the three aromatic rings in the starting material was the correct structure did not preclude the process of making alizarin from anthracene to be patented. Ideas of synthetic transformation were also maturing in the late 19th century and the patented oxidative route to alizarin from anthracene was the reverse type of process to the reductive procedures used to convert the natural product into anthracene.

By the end of the 19th century a range of natural products had been chemically synthesised including indigo (several routes: 1878, 1880, 1896), glucose ($C_6H_{12}O_6$, 1887) and coniine ($C_8H_{17}N$), the toxic active constituent of hemlock in 1886 (Figure 4.28). Organic synthesis as a discipline was evolving rapidly at the start of the 20th century and many new reactions were being discovered through experimental work in academic and industrial laboratories. Reactions became more reproducible and reliable with the development of chemical glassware and the structural identity of compounds was advanced with improvements in analytical techniques and instrumentation. The new theories and ideas which emerged from experimental work led to a better understanding of bonding and reactivity of compounds, and synthetic routes to target molecules are now firmly based on rational design processes.

Some compounds in organic chemistry are recognised as milestone molecules because of the profound changes in the subject which developed as a consequence of their synthesis. Perkin's mauve was one of these compounds and in 1899 acetylsalicylic acid (aspirin) became another as this was a designed molecule and triggered the birth of the pharmaceutical industry. Perkin's discoveries gave him and his family financial security and in 1874 he sold his company and returned to pure research in chemistry. He published widely in areas ranging from the synthesis of coumarin,

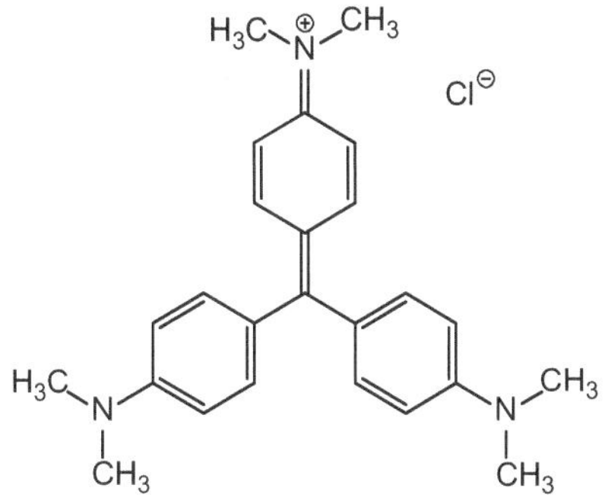

Gentian violet

FIGURE 4.29 Structure of gentian violet.

which formed the basis of the perfume industry, to the magneto-optical Faraday effect which is used in modern high-density data storage devices. Echoes of Perkin's synthetic dye discoveries persist to this day in biochemistry laboratories around the world as hexamethyl pararosaniline chloride (crystal violet or gentian violet, Figure 4.29) is a standard histological stain to differentiate between Gram-positive and Gram-negative bacteria.

4.5 CONSERVATION AND RESTORATION OF HISTORIC COLOURED ARTEFACTS

Works of art change over time; exposure to light can cause once vibrant colours to fade, cracks and flaking of material can occur as a result of changes in temperature and humidity and dust particles can settle and become trapped in a textured surface. Whether a particular painting or other historic artefact is cleaned, stabilised and repaired to prevent further deterioration (conservation) or restored to how it looked when new (restoration) depends on many factors such as the cultural significance of the object and cost. Conservation and restoration procedures have evolved over the years as new techniques have developed but understanding the chemistry between the materials used to make an artwork and the outside world is central in identifying a strategy for its preservation.

In many modern conservation projects of historic artefacts the first step involves analysing the object using non-invasive and non-destructive spectroscopic techniques. For example, X-ray fluorescence uses a focused beam of X-rays to identify the elements present in a sampled area of an art work and this technique was also used to obtain geochemical signatures of the sarsen and bluestones of Stonehenge (see Section 3.7.4). Scanning a painting using infrared reflectography can detect features beneath the paint such as the presence of an underdrawing, the initial preparatory sketch for the composition. If the artist used a material such as charcoal, which absorbs rather than reflects the infrared radiation, the underdrawing appears dark in the black-and-white reflectogram image. This technique can also uncover hidden images, changes made by the artist during the process of painting and can be instrumental in determining whether a painting is an original or a copy. Information on the painting's layer structure and constituent parts often involves an invasive procedure in which a miniscule sample (typically ~0.25 mm^2) is taken from either an area of damage or at the very edge of painting and hidden by the frame. Sampling and non-invasive techniques were both involved in the analysis and subsequent restoration of the painting *Portrait of Alexander Mornauer.*

When the 15th-century painting *Portrait of Alexander Mornauer* (Figure 4.30) was acquired by the UK National Gallery in 1990 it became apparent that some features such as the texture and colour of blue background paint were unusual for a picture from that era. Analysis of a small sample of the background found that the blue was Prussian blue which had not been invented until early in the 18th century and that the medium in this overpaint layer was poppyseed oil whilst the original paint was bound in linseed oil. Furthermore, the underdrawing showed a taller hat on the figure and this design of hat was revealed to be in the original picture when the overpaint was removed. The overpainting and changes to the original picture had been done somewhere between around 1720 and the late 18th century when it was in the Stowe collection in Buckinghamshire. The changes may have been made to suit changing tastes of the time or could have been a deliberate attempt to heighten the resemblance to the work of the 16th-century portraitist Hans Holbein the Younger, thereby making the painting more appealing to an art collector in addition to increasing its value.

FIGURE 4.30 *Portrait of Alexander Mornauer.* The left image is the *Portrait of Alexander Mornauer* before cleaning and shows a blue background and smaller hat. The right-hand image shows the picture after cleaning and restoration. (Photocredit – © The National Gallery, London.)

4.5.1 COLOUR CHANGES IN PAINTINGS

Many pigments in decorative artworks change over time; some will fade due to light damage whilst others may react with substances in the environment. Photodegradation causes yellow orpiment pigment (As_2S_3), for example, to become paler due to the formation of white arsenic oxides. The oxides are slightly more water-soluble and the pigment can migrate into surrounding parts of a painting. The ageing process also leads to the darkening of nearby copper- and lead-containing pigments by formation of copper and lead sulphides due to a reaction with volatile sulphides such as hydrogen sulphide (H_2S) emitted upon decomposition of the orpiment paint. The red arsenic pigment realgar (α-As_4S_4) photo-ages to the orange-yellow colour of pararealgar (β-As_4S_4), a compound with the same formula as realgar but with a different structural arrangement of sulphur and arsenic atoms (Figure 4.31).

Recent work on some of Vincent van Gogh's paintings of the Provencal period (1888–1890) has given insight into why the vivid violet colour of the iris flowers have faded to a more distinctly blue colour. Van Gogh himself noticed that paintings in which he had used this violet colour were fading and he tried to overcome this problem by working with thicker brushstrokes. Van Gogh created the violet colour by mixing blue pigments (typically cobalt blue or Prussian blue) with red geranium lake pigments. Geranium lake is based on eosin Y, a vibrant red water-soluble dye, which was converted into an insoluble pigment by complexing it with an aluminium mordant. The fading phenomenon occurs because of the photosensitivity of the eosin chromophore in the geranium lake. Light causes eosin Y to abstract a hydrogen radical from fatty acids in the oil binder, thus forming a semi-reduced eosin Y radical species in which the conjugated π-system, the source of the colour, has been broken. The radical can re-form eosin Y by loss of a hydrogen radical or react with atmospheric oxygen and break down into two colourless compounds: 2-(3,5-dibromo-2,4-dihydroxybenzoyl) benzoic acid (BHBA) and 2,6-dibromobenzene-1,4-diol (DBB) (Figure 4.32).

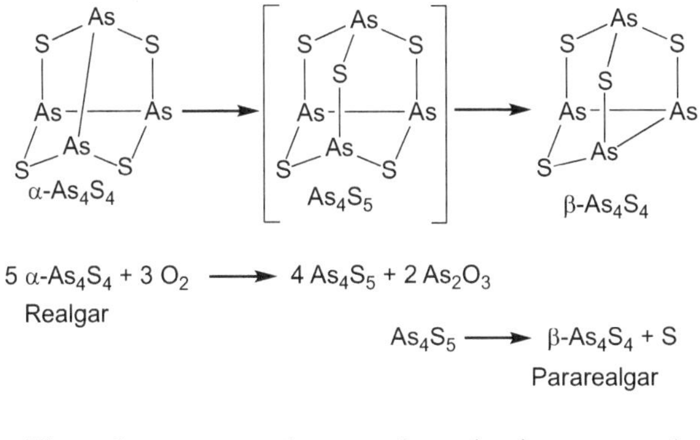

$$5\,\alpha\text{-}As_4S_4 + 3\,O_2 \longrightarrow 4\,As_4S_5 + 2\,As_2O_3$$

Realgar

$$As_4S_5 \longrightarrow \beta\text{-}As_4S_4 + S$$

Pararealgar

FIGURE 4.31 The realgar to pararealgar transformation is a structural modification in which the effect of light and oxygen causes a sulphur atom to be inserted between two arsenic atoms of the α-As_4S_4 cage structure to form an intermediate As_4S_5 species and arsenic oxide (As_2O_3). This intermediate decomposes into β-As_4S_4 by breaking an As–S–As bond and releasing a sulphur atom. The conversion then continues with the released sulphur atom reacting with another molecule of realgar to produce a molecule of pararealgar etc.

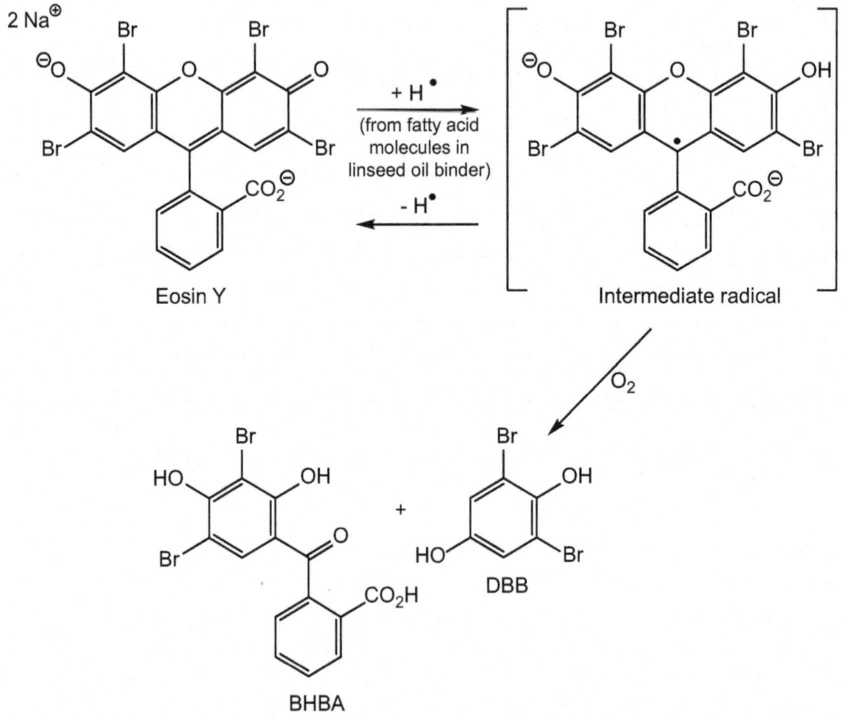

FIGURE 4.32 Proposed photodegradation pathway of the eosin-based geranium lake pigment. These findings were the result of experiments involving monitoring reference paint layers made from geranium lake pigments prepared using late 19th century recipes mixed with linseed oil. The samples were analysed by a chromatography-mass spectrometry system to characterise degradation products over a period of one year.

Geranium lake was the only bromine-containing lake pigment available to van Gogh and scanning paintings with an X-ray fluorescence analyser can give a chemical map of where the violet pigment was used. Using information from a broad range of analytical techniques some of van Gogh's paintings have been digitally reconstructed to show how the original colours would have appeared. Furthermore, determining how pigments and dyes change over time can help to establish the appropriate conditions for storing and exhibiting coloured art works in museums and galleries.

4.5.2 THE BAYEUX TAPESTRY

The Bayeux Tapestry is an embroidery of dyed woollen threads on a linen cloth which dates from the end of the 11th century.[7] This medieval textile depicts scenes leading up to the Norman invasion of England and the Battle of Hastings and measures almost 70 m long by about 50 cm high. The embroidered scenes show daily life in addition to the military material and as such it contains a wealth of historical information that is rarely seen in works of art which have survived from the medieval period.

Bishop Odo of Bayeux, half-brother of William the Conqueror, is believed to have commissioned the tapestry which may have been intended for propaganda purposes to justify the invasion to the general population. The embroidery work itself is believed to have been done by skilled English needleworkers in Canterbury in the 1070s. A replica of the tapestry was created in 1885–1886 by a group of 35 embroiderers from the Leek Embroidery Society and is on permanent display in the Reading Museum, Berkshire. Both the original Bayeux Tapestry and the Reading replica can be explored electronically online (see *Further Reading* for details).

A backing strip of linen has been in place on the reverse side of the Bayeux Tapestry for several centuries to give extra support and protection. Temporary removal of this material in conservation work undertaken in 1982–1983 enabled the reverse side of the embroidered work to be studied in detail and for samples of wool to be removed for analysis. With little direct exposure to light, the threads on the back were more colourful and better preserved compared to those on the story-telling front. Furthermore, examination of the exposed stitching on the reverse side has also given insight into the tapestry's construction and the working practices of the needleworkers. For example, the back of the embroidery stitching from 19th century repairs was very untidy compared to the original Anglo-Saxon work, suggesting that the later needleworkers were less well-trained that their earlier counterparts.

Extraction of dye from the woollen thread samples followed by chromatographic analysis of the isolated material identified that all the colours in the original tapestry could be obtained from just the red, yellow and blue dyes from madder, weld and indigo plants respectively. The 1980s work also found that the greater degree of fading in the embroidered designs repaired in the late 19th century was due to the use of wool coloured with synthetic aniline-based dyes. Synthetic dyes manufactured in the 1850s and 1860s were relatively inexpensive and available in a range of bright colours and so were an attractive choice for the 19th-century restorers of the tapestry although some of these compounds were known to be more light-sensitive than traditional natural dyes.

More precise identification of the synthetic and original natural dyes has been obtained from recent fibre-optics reflectance spectroscopy (FORS) and hyperspectral imaging (HIS) of the embroidery. This is a non-invasive procedure performed in situ without any contact with the historic textile, in contrast to the sampling techniques used in the 1980s work on the tapestry. Visible reflectance spectra of the tapestry essentially yield dye signatures for the different coloured embroidery threads which can be compared with those obtained on reference textile samples from the 1870s. Various forms of rosaniline blue and several derivatives, aurin (red) and methylaniline violet, were detected along with plant-based dyes, indicating that the restorers used threads coloured with both natural and synthetic dyes. Identification of methylaniline violet as a component of the black threads suggests that the repair work was carried out no earlier than 1866, the year this dye was first manufactured, whilst photographs taken of the restored tapestry in 1872 provide an end date for the repairs.

Establishing the chemical identity of natural dyes in historic textiles through FORS and other spectroscopic techniques also offers the opportunity to create dyed wool or other fibres which closely match those originally used in ancient artworks. Freshly dyed samples of wool are however brightly coloured whereas in historic textiles the colours could have faded due to environment-induced ageing. Exposure of newly coloured fibres to high-intensity light can achieve artificial ageing and give materials for conservation work that closely match those in the historic work. Reproductions of historic textiles using freshly coloured materials can, in the same manner as with classic paintings, reveal what the original object looked like when newly completed and help people connect with the past.

4.6 STRUCTURAL COLOUR AND DRAGONFLIES

Most colouration in animals and plants is based on the selective absorption of light by molecular pigments. However, colour can also arise from the interaction of light with physical structures within the outermost parts of an organism and the effect can be quite striking; colours can be iridescent and they often have a metallic appearance. Structural colours are most common in invertebrates and are produced, for example, by the shells of certain molluscs and the wings and/or bodies of some *Lepidoptera* (butterflies and moths) and *Odonata* (dragonflies and damselflies) species. The males of both the *Calopteryx* species found in Britain (Figure 4.33) have blue-green irides-cence wings which makes these insects very conspicuous as they engage in activities such as territorial disputes and courtship displays. The extent of the structural colou-ration differs between the two species with *C. splendens* (banded demoiselle) males having an iridescent patch in each wing compared to *C. virgo* (beautiful demoiselle) males whose wings are almost completely iridescent.

Odonata wings are a composite of chitin, a polymeric material similar to cellu-lose, and protein which gives a lightweight but strong structure. Interspersed within layers of this chitin polymer matrix in the *Calopteryx* damselflies is melanin, a brown pigment. Layers of chitin and melanised chitin form a nanostructure with internal periodic features which have the same order of magnitude as the wavelength of light, and it is the interference of light with this internal structure which causes the iridescence colours. Viewing *C. splendens* wings under transmitted light conditions demonstrates that although they are pigmented brown it is their physical nanostruc-ture that makes them appear an iridescent blue (Figure 4.34a). Whilst iridescence is

(a)   (b)

FIGURE 4.33   (a) Calopteryx splendens male. (b) Calopteryx virgo male.

used advantageously by the *Calopteryx* damselflies for inter- and intraspecies communication, it has its drawbacks as these insects are also highly visible to predators and if caught can become food for birds or larger species of dragonflies. The wings of the insect are however indigestible to many predators and may be discarded when the insect becomes a meal. These discarded remains can be collected and, as the iridescent effect survives the death of the insect, they can be used for further analysis (Figure 4.34b).

The melanised and non-melanised chitin nanostructure functions as a multilayer reflector stack within the insect's cuticle. Such a structure consists of alternating layers of high and low refractive index material and a certain proportion of the incident light is reflected from each interface (Figure 4.35). Reflections from successive interfaces can interfere constructively or destructively with each other when they emerge from the stack. For the interference to be constructive the reflections must emerge in the same phase and for this to occur the optical path length difference between successive reflections is a half-integral multiple of the wavelength and given by the equation

$$2nd\cos\theta = (m + \tfrac{1}{2})\lambda$$

The colour reflected therefore depends on the optical thickness [the product (nd) of the refractive index and the actual thickness] of the layers and the angle of incidence at the boundary interface. As the angle of incidence changes, different wavelengths constructively interfere; oblique angles of incidence give a reflected colour at a shorter wavelength compared to that for light at normal incidence. In a multilayer system the direction of the propagation of the wave within the two different media is different and the system is no longer ideal which complicates the mathematics. However, by considering a single film reflector system in air, the above equation with $m = 0$ becomes $\lambda = 4nd\cos\theta$. The value of $\cos\theta$ decreases from 1 as θ increases from zero degrees (normal incidence) which shifts the reflected light in the direction

(a)   (b)

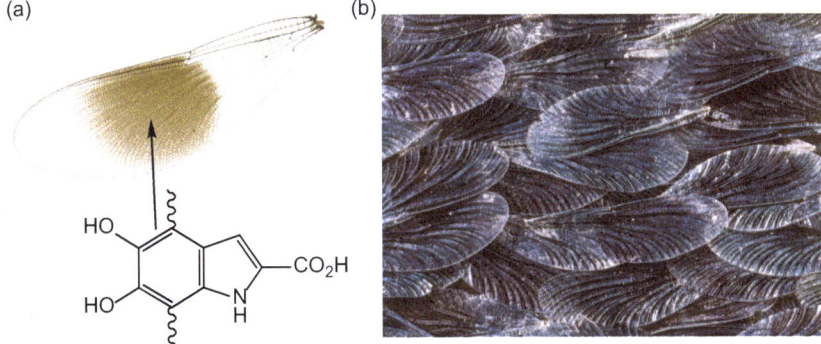

Melanin, a polymer of 5,6 dihydroxyindole-2-carboxylic acid

FIGURE 4.34 Male *C. splendens* wings collected in the field and viewed under (a) transmitted and (b) reflected light.

1. Light arriving at the multilayer reflector system is partially reflected and partially refracted at the top surface. The refracted light continues to pass through the top layer and when it reaches the bottom surface of this layer, and crosses into a layer with a lower refractive index, a further portion is partially reflected and partically refracted. This process of reflection and refraction continues as the light passes through the layers of the nanostructure.

2. Light waves undergo a 180° phase change when reflected from the surface with a higher refractive index than that of the medium in which they are travelling, i.e. at the low to high interfaces.

3. Light waves slow down in the higher refractive index layer. Light reflected from the lower interface of this layer must be retarded by a half-wavelength to enable the reflected beams from the top and bottom interfaces to be in phase with each other when they emerge from the nanostructure.

As light passes twice through each layer the optical path length (refractive index, n x actual thickness, d) must therefore be **one-quarter wavelength** for maximum constructive interference when light is reflected normally, i.e. perpendicular to the surface.

4. Reflected waves emerge in phase from the stack and constructively interfere. If the spacing of the layers approaches one-quarter of the wavelength of visible light then the reflected light is coloured.

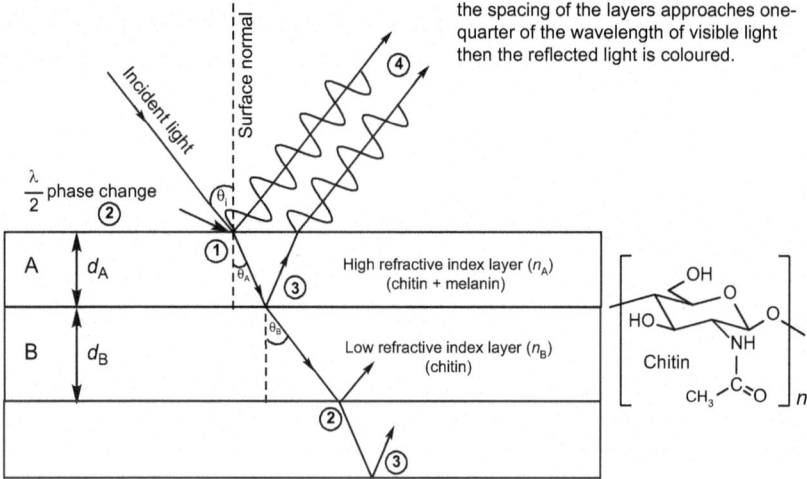

FIGURE 4.35 Diagram of a multilayer interference reflector. An ideal reflector has every layer in the stack equal to a ¼ wavelength of light, i.e. $n_A \times d_A = n_B \times d_B = \lambda/4$. A mathematically non-ideal system is one in which $n_A \times d_A \neq n_B \times d_B$ but in which the sum of the optical thicknesses of each layer-pair $[(n_A \times d_A) + (n_B \times d_B)]$ is equal to one-half of the wavelength. In nature non-ideal reflectors can achieve the desired biological function for the host organism despite not being the 'ideal' quarter-wave stack. Thin-film structures with layered thicknesses designed to produce a destructive interference effect can be applied to the surfaces of optical devices such as spectacles, camera lenses and solar cells to reduce the amount of light reflected off the surface.

of shorter and bluer wavelengths. Shifting patterns of brightness and colour with the angle of observation is the definition of iridescence. Very high levels of reflectivity occur when the stack has 10 or more high refractive index layers which confers the biological surface with a highly metallic appearance, another distinctive feature of iridescence.

Plants also use structural colour in their tissues for display and defence. Iridescence in flowers may enhance pigment-based colour and aid pollination whilst multilayer reflectors in the leaves of shade-dwelling plants can provide photoprotection during periods when exposure to harmful levels of UV radiation may be damaging. Physical nanostructures in plants and animals, as with the chemical structure of dyes and pigments, provide inspiration for the design of new coloured materials. A reflective blue paint finish of some cars was made from multiple layers of titanium and hafnium oxides and was the culmination of research to mimic the wing nanostructure of the *Morpho menelaus* butterfly. Nanostructures in the waxy coatings of blueberries give these fruits their colour, and this colour could be reproduced in the lab by an extraction and recrystallisation procedure. This finding offers a potentially new approach for the sustainable manufacture of nanoparticle colourants and coatings in the future.

NOTES

1   In quantum mechanics light can be viewed as being composed of individual particles called photons or as a wave characterised by its frequency and wavelength. Molecular quantum mechanics is a quantitative theory to explain properties of atoms and molecules and requires an advanced knowledge of mathematics to solve the complex expressions. However, solutions to the complex equations can be interpreted in qualitative terms to give a visual picture of atomic structure and particularly the spatial distribution of electrons which is important to bonding. Quantum mechanics gives a more comprehensive picture of atomic structure than the earlier Bohr planetary model and is currently the most successful way to explain properties of atoms.

2   Cisplatin is an anticancer agent. The formula of the drug administered is not the form in which the drug is active. The active form has at least one chloride ligand replaced by water. In plasma, the high chloride ion concentration prevents the hydrolysis of cisplatin, but when it enters the cell the much lower chloride concentration prompts its reaction with water to give the positively charged species $[PtCl(H_2O)(NH_3)_2]^+$ and $[Pt(H_2O)_2(NH_3)_2]^{2+}$. These active forms can bind to guanine bases in DNA resulting in intra- and inter-strand crosslinks which deform the DNA structure thereby interfering with DNA replication mechanisms which can lead to cell death. The cytotoxic effects of cisplatin were discovered by accident in the 1960s during a study of the effects of electric currents on cells; *Escherichia coli* cells were found to form long filaments but they did not divide. Further research showed that inhibition of bacterial cell division was due to cisplatin, generated from the platinum electrodes and the ammonium chloride present in the medium.

3   An inhibitory rather than an excitatory signal is a somewhat unusual feature of information handling in the retina compared to the situation in most other sensory neurons. Although this may appear odd, the visual system relies on a consistent relationship between changes in the number of open cation channels and the rate of neurotransmitter release.

4   Converting a chance observation into the discovery of a new and useful material or into a breakthrough in the understanding of science requires curiosity, luck and sagacity or as Louis Pasteur phrased it, '*Dans les champs de l'observation, le hasard ne favorise que les esprits préparés*' or 'In the fields of observation, chance favours only the prepared mind'.

5   One of the early syntheses of indigo involved oxidation of naphthalene ($C_{10}H_8$ a bicyclic aromatic hydrocarbon derived from coal tar) to phthalic anhydride using fuming sulphuric acid. This was a slow step and the synthesis only became economically viable from a serendipitous accident when one of the chemists monitoring the oxidation reaction broke a mercury thermometer and the mercury fell into the reaction mixture. In the acidic oxidising environment, the mercury was quickly converted to mercury(II) sulphate, $HgSO_4$, a hitherto unknown but effective catalyst for the oxidation of naphthalene.

6 Polyacrylonitrile is usually a copolymer with acrylonitrile as the main monomer and a small amount of another compound such as acrylic acid or sodium allylsulphonate which gives an overall negative charge.

7 The designs in the Bayeux Tapestry are sewn directly onto the fabric using a needle and thread and so it is more accurate to describe the work as an embroidery. However, 'Bayeux Tapestry' has remained in common usage although in a true tapestry the design is woven into the fabric as it is created on a loom.

FURTHER READING

Most of the pigments discussed in this chapter are readily available from art suppliers. However, some contain toxic heavy metals such as lead, mercury and cadmium and therefore must be used with extreme care. Several of the experiments in Chapter 6 involve natural dyes and gold nanoparticles and additional information and reference material on these compounds can be found therein.

BOOKS

Bomford, D., Dunkerton, J. and Wyld, M. (2009) *A Closer Look: Conservation of Paintings*. London: National Gallery.

Burrows, A., Holman, J., Lancaster, S., Overton, T., Parsons, A., Pilling, G. and Price, G. (2021) *Chemistry³: Introducing Inorganic, Organic and Physical Chemistry*. 4th Ed. Oxford: Oxford University Press.

Delamare, F. and Guineau, B. (2000) *Colour: Making and Using Dyes and Pigments*. London: Thames & Hudson.

Garfield, S. (2000) *Mauve – How One Man Invented a Colour That Changed the World*. London: Faber and Faber.

McCleverty, J. (1999) *Chemistry of the First-Row Transition Metals*. Oxford: Oxford University Press.

O'Shea, M. (2005) *The Brain: A Very Short Introduction*. Oxford: Oxford University Press.

Walmsley, I. (2015) *Light: A Very Short Introduction*. Oxford: Oxford University Press.

Winter, M. J. (2016) *Chemical Bonding*. 2nd Ed. Oxford: Oxford University Press.

Wieseman, M. E. (2010) *A Closer Look: Deceptions and Discoveries*. London: National Gallery.

ARTICLES

Budd, J., Miller, B. S., Weckman, N. E., Cherkaoui, D., Huang, D., Thomas Decruz, A., Fongwen, N., Han, G.-R., Broto, M., Estcourt, C. S., Gibbs, J., Pillay, D., Sonnenberg, P., Meurant, R., Thomas, M. R., Keegan, N., Stevens, M. M., Nastouli, E., Topol, E. J., Johnson, A. M., Shahmanesh, M., Ozcan, A., Collins, J. J., Fernandez Suarez, M., Rodriguez, B., Peeling R. W. and McKendry, R. A. (2023) Lateral Flow Test Engineering and Lessons Learned From COVID-19. *Nature Reviews Bioengineering*. [Online] 1. pp. 13–31. Available from: https://www.nature.com/articles/s44222-022-00007-3. [Accessed: 13th March 2024].

Centeno, S. A., Hale, C., Carò, F., Cesaratto, A., Shibayama, N., Delaney, J., Dooley, K., van der Snickt, G., Janssens, K. and Stein, S. A. (2017) Van Gogh's *Irises* and *Roses*: The Contribution of Chemical Analysis and Imaging to the Assessment of Color Changes in the Red Lake Pigments. *Heritage Science*. [Online] 5. Article no. 18. Available from: https://doi.org/10.1186/s40494-017-0131-8. [Accessed: 21st January 2025].

Chavanne, C., Verney, A., Paquier-Berthelot, C., Bostal, M., Buléon, P. and Walter, P. (2022) Bayeux Tapestry: First Use of Early Synthetic Dyes for the Restoration of a Masterpiece. *Dyes and Pigments*. [Online] 208. Article no. 110798. Available from: https://doi.org/10.1016/j.dyepig.2022.110798. [Accessed: 19th February 2024].

Cova, T. F. G. G., Pais, A. A. C. C. and Seixas de Melo, J. S. (2017) Reconstructing the Historical Synthesis of Mauveine from Perkin and Caro: Procedure and Details. *Scientific Reports*. [Online] 7. Article no. 6806. Available from: https://doi.org/10.1038/s41598-017-07239-z. [Accessed: 3rd November 2019].

Kefalov, V. J. (2012) Rod and Cone Visual Pigments and Phototransduction through Pharmacological, Genetic, and Physiological Approaches. *Journal of Biological Chemistry*. [Online] 287 (3). pp. 1635–1641. Available from: https://doi.org/10.1074/jbc.R111.303008. [Accessed: 15th February 2024].

Middleton, R., Tunstad, S. A., Knapp, A., Winters, S., McCallum, S. and Whitney, H. (2024) Self-Assembled, Disordered Structural Color from Fruit Wax Bloom. *Science Advances*. [Online] 10 (6). Article no. eadk4219. Available from: https://doi.org/10.1126/sciadv.adk4219. [Accessed: 26th May 2024].

Nicolaou, K. C. (2014) Organic Synthesis: The Art and Science of Replicating the Molecules of Living Nature and Creating Others Like Them in the Laboratory. *Proceedings of the Royal Society A*. [Online] 470. Article no. 20130690. Available from: https://doi.org/10.1098/rspa.2013.0690. [Accessed: 30th April 2024].

Parker, A. R. (2000) 515 Million Years of Structural Colour. *Journal of Optics A: Pure and Applied Optics*. [Online] 2 (6). pp. R15–R28. Available from: https://doi.org/10.1088/1464-4258/2/6/201. [Accessed: 27th May 2024].

Tsin, A., Betts-Obregon, B. and Grigsby, J. (2018) Visual Cycle Proteins: Structure, Function, and Roles in Human Retinal Disease. *Journal of Biological Chemistry*. [Online] 293 (34). pp. 13016–13021. Available from: https://doi.org/10.1074/jbc.AW118.003228. [Accessed: 15th February 2024].

Vukusic, P., Wootton, R. J. and Sambles, J. R. (2004) Remarkable Iridescence in the Hindwings of the Damselfly *Neurobasis chinensis chinensis* (Linnaeus) (Zygoptera: Calopterygidae). *Proceedings of the Royal Society London B*. [Online] 271. pp. 595–601. Available from: https://doi.org/10.1098/rspb.2003.2595. [Accessed: 26th May 2024].

Wentrup, C. (2022) Origins of Organic Chemistry and Organic Synthesis. *European Journal of Organic Chemistry*. [Online] 2022 (25). Article no. e202101492. Available from: https://doi.org/10.1002/ejoc.202101492. [Accessed: 29th April 2024].

WEBSITES

Bayeux museum (2024) *Explore the Bayeux Tapestry Online.* [Online] Available from: https://www.bayeuxmuseum.com/en/the-bayeux-tapestry/discover-the-bayeux-tapestry/explore-online/. [Accessed: 15th May 2024].

Reading museum (2024) *Britain's Bayeux Tapestry.* [Online] Available from: https://www.readingmuseum.org.uk/collections/britains-bayeux-tapestry. [Accessed 28th May 2024].

5 Communication

5.1 INTRODUCTION

Communication is in essence the transmission of information between a sender and a recipient. A cue, which can take a variety of formats, is dispatched by the sender and is recognised, and then interpreted, by the recipient. Recall that purple coloured clothing was worn by some individuals in ancient civilisations to signal their high status to the general public (Section 4.3.3.1) and that iridescent wings are used by male *Calopteryx* species of damselfly to send information to competitors or potential mates (Section 4.6). Many insects use chemical scents or pheromones to carry signals to other member of their species and this form of olfactory communication is discussed in further detail in Section 5.2 of this chapter.

Methods of communication within the human body involve complex and interconnected networks which regulate many physiological processes. Earlier chapters considered how information flows in the visual system from the eye to the brain, how molecular mechanisms underlie pain transmission and why a breakdown in the onward flow of information relating to blood glucose levels has health implications. Salient features of these biological communication systems are summarised as follows. Blood glucose levels are not constant; they increase following a meal or decrease during a period of fasting, and this change is detected by cells in the pancreas which triggers the secretion of the hormones insulin or glucagon as required. A fault in the information flow from initial cue to hormone response results in diabetes. Treatment of this condition (Section 2.7) uses protocols to gather information on blood glucose levels and to generate a response to mimic that provided by healthy pancreatic cells. G protein-coupled receptors (GPCRs) are molecular sensors which bridge input signals and cellular responses in the visual system and pain pathways. Vision starts with the absorption of photons by the opsin family of light-sensitive GPCRs and the light energy causes a change in the structure of the receptor protein (Section 4.2.3). This structural change acts as a message for the initiation of cellular processes, the output of which is translated into electrical signals that ultimately reach the brain. Acute pain can be treated by the administration of analgesics such as morphine (Section 2.4.6). When morphine binds to a G protein-coupled opioid receptor it triggers a cellular cascade leading to the inhibition of nerve impulses, an observable sign of which is relief from a distressing sensation.

Exchange of messages between humans often takes place using language, through either speech, writing or gestures. The origin of language lies deep in human prehistory, although inferences from the fossil record and anatomical evidence from archaeological specimens generally suggest a date somewhere between 1.4 million and about 600,000 years ago, i.e. during the *Homo erectus* and *Homo heidelbergensis* periods of evolution. The earliest known evidence of writing is in the form of Sumerian clay pictograms from Mesopotamia (modern-day Southern Iraq) just over 5,000 years ago.[1] However, Egyptian hieroglyphic script is considered to be the first true writing

DOI: 10.1201/9781003562313-5

system as some glyphs signify sounds and this is therefore a written form of a spoken language. Hieroglyphs date from around 3,000 BCE and were largely carved into the stone of important religious and royal buildings but they had fallen out of use by about the 4th century CE. Hieroglyphic writing evolved into a less pictorial and more abbreviated cursive form known as Demotic script. This script enabled faster writing with reed pens on papyrus and was used in administrative documents from about the 7th century BCE to 5th century CE. A more alphabetical script system, incorporating elements of the ancient Greek language, developed in Egypt as the Ptolemies of Greco-Macedonian descent took power between 305 and 30 BCE. Three versions of a message written in Egyptian hieroglyphs and Demotic and ancient Greek scripts from 196 BCE were discovered on a single black granite stone slab in the late 18th century near Rashid (Rosetta), a port town of the Nile Delta. The message on the Rosetta Stone written in the known language of ancient Greek was the key to deciphering the long-forgotten ancient Egyptian languages. The discovery that some of the glyphs represented phonetic sounds in royal names established Egyptian hieroglyphs as the first written language and as a result the invention of writing is taken to be 3,100 BCE. This date formally marks the transition from prehistory, the time before writing was invented, to history and the era of documentation of past events and their interpretation.

Papyrus, made from the plant *Cyperus papyrus* native to the Egyptian Nile Valley, was the most commonly used writing surface in antiquity. Over time there was a progression towards parchment and vellum from animal skins which were better suited to damper climates, and thence to paper which was cheaper but less durable. The black and red inks used by the ancient Egyptians for writing were made primarily from soot and red ochre (haematite, Fe_2O_3), and mixed with a binder and suspended in water. Ink composition changed with the advent of the printing press in the 15th century and then again with modern commercial printing and photocopying. Writing and the printing press both enabled ideas and knowledge to spread more easily and become more accessible to the wider population. This trend is continuing with smartphone technology and people are now able to communicate instantaneously with others anywhere in the world. Photo messaging is a popular adjunct to the original verbal and text-based ways of communicating on a smartphone and many people use the phone camera for all their digital photography.

The 21st century information revolution is changing the way people work, interact with others and live their lives and the magnitude of the change is causing a cultural shift in society. The role of chemistry in this modern revolution comes about through the materials and processes used to make the electronic devices. Firstly, electrochemistry is at the heart of battery technology and secondly, self-organising molecules which interact with light in interesting ways when an electric field is applied are integral to the operation of many information displays. As the modern communication revolution continues to unfold it is pertinent to take a closer look at the chemistry of batteries and screens in the current generation of electronic devices (Sections 5.4 and 5.3 respectively).

Cameras on digital devices are making it easier for more people to take and share photographs and the role of photographs in society is changing as a result. Traditional photographs taken with a film or analogue camera are predominantly a means of remembering and sharing experiences with families and a small group of friends.

Digital photographs still fulfil this function, but additionally the captured image can become a message to be shared more widely through the distribution power of the internet. The invention of photography in the 19th century started the process of democratising the capturing, sharing and communication of visual images and the chemical magic of the analogue photographic image is the focus of the final section of this chapter. Photographs from the 19th and 20th centuries are the visual traces of past lives, events, landscapes and townscapes and they can be appreciated for their aesthetic qualities in the same manner as paintings. Moreover, certain photographs may be primary data sources and their analysis can give historians insight into the social and cultural influences current at the time the image was taken.

5.2 INSECT PHEROMONES: COMMUNICATION BY SMELL

Insects use a combination of visual and auditory signals, body movements and phero-mones (or chemical cues) to communicate with other members of their own species. Pheromones are released externally from the glands of an insect as a liquid which evaporates into the surrounding air. The now gaseous molecules travel by diffusion through the air and are detected by receptors on the antennae of the receiver insect. Chemoreceptors on the antennae are very sensitive, and highly specific to a particu-lar insect scent, and so need only capture a few molecules of the pheromone in order to trigger a behavioural response. The nature of the communication with pheromones requires these molecules to be volatile organic molecules which correlates with a low boiling point, and as a result they usually have molecular weights which do not exceed 250 g/mol. Many of the known insect sex attractants, which are the most widely studied group of pheromones, are carbonyl compounds or carboxylic acid derivatives with long chains of carbon atoms containing one or more double bonds (Figure 5.1a). Sex pheromones are often a blend of two or more compounds in a precise ratio in order to create a signal that is unique to a particular species. Social insects such as ants and bees also use alarm and trail pheromones to signal to other members of their colony of imminent danger or to guide them towards new food sources (Figure 5.1b and c respectively).

Interfering with an insect's ability to send or receive chemical signals is a biological method of monitoring or controlling insect populations. Traps baited with synthetic sex pheromones of the pine beauty moth (*Panolis flammea*) were used to this effect and the chemistry of these compounds is the focus of Section 5.2.2. A prerequisite for work of this nature is identification of the compounds in the sex pheromone. In pio-neering work on the silkworm moth *Bombyx mori* between 1939 and 1959, the phero-mone was extracted from the abdomens of 500,000 female moths to give just 6.4 mg of a single compound which was identified as (10*E*,12*Z*)-hexadeca-10,12-dien-1-ol. Male silkworm moths responded behaviourally to bombykol at concentrations of 3,000 molecules/cm^3 of air, i.e. a moth's antennae are nearly as sensitive as a dog's nose.

Chemical signalling is the primary source of communication in night-flying moths and it is commonly the females which are the producers of pheromones. In contrast it is the males of their diurnal relatives, the butterflies, which emit the chemical scent and these insects use a combination of olfactory and visual cues for communication.

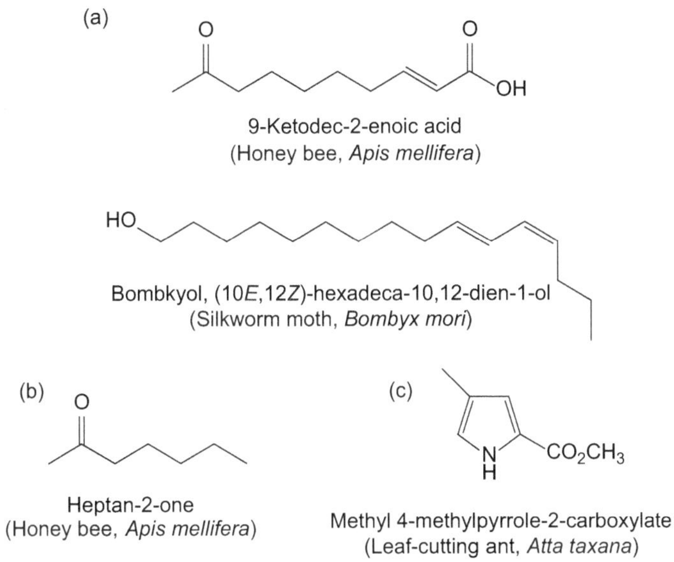

(a)

9-Ketodec-2-enoic acid
(Honey bee, *Apis mellifera*)

Bombkyol, (10*E*,12*Z*)-hexadeca-10,12-dien-1-ol
(Silkworm moth, *Bombyx mori*)

(b)

Heptan-2-one
(Honey bee, *Apis mellifera*)

(c)

Methyl 4-methylpyrrole-2-carboxylate
(Leaf-cutting ant, *Atta taxana*)

FIGURE 5.1 Molecular structures of pheromones from selected species: (a) sex pheromones, (b) an alarm pheromone and (c) a trail pheromone.

Investigations into the sex pheromones of pierid butterflies has identified both aphrodisiac and anti-aphrodisiac compounds in their chemical scents and this has given insight into some of the subtler aspects of their reproductive behaviour (Section 5.2.1).

5.2.1 APHRODISIAC AND ANTI-APHRODISIAC COMPOUNDS

Green-veined white (*Pieris napi*) and large white butterflies (*Pieris brassicae*) are members of the Pieridae family of butterflies as this group all lay skittle-shaped eggs. These two species of butterfly can be confused, especially when they are in flight, as both are predominantly white with black markings (Figure 5.2), but when they settle on vegetation the green veins on the underside of the wings of *P. napi* are easily seen. The large white and the closely related small white butterfly (*Pieris rapae*) are sometimes collectively referred to as cabbage whites, as the larvae (or caterpillars) of these two species are the only ones which cause damage to crops in the UK; having a strong preference for cultivated varieties of cabbage and brussels sprouts. The green-veined white caterpillars in contrast feed on wild cruciferous plants such as garlic, mustard and cuckoo flower.

Mate location in butterflies initially occurs visually and this is followed by the release of short-range male aphrodisiac pheromones from scent glands on their wings. The aphrodisiac pheromone of *P. napi* male butterflies is citral, a 1:1 mixture of neral and geranial, whilst *P. brassicae* males use an aphrodisiac blend consisting of brassicalactone together with phytol and 6,10,14-trimethylpentadecan-2-one (Figure 5.3). Isolation and identification of the chemical scents from the male pierid

FIGURE 5.2 Mating pairs of (a) *P. brassicae* (large white) and (b) *P. napi* (green-veined white) butterflies.

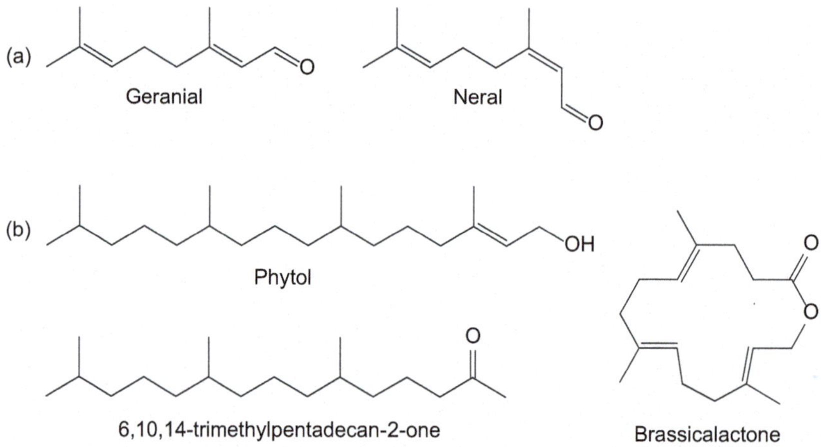

FIGURE 5.3 Aphrodisiac compounds found in male pierid species of butterfly. (a) *P. napi*: citral, an equal mixture of geranial and neral and (b) *P. brassicae*, a three-component blend of a cyclic ester, a linear alcohol with 20 carbon atoms and an 18-carbon long ketone.

wings involved far fewer insects than in the bombykol work as modern instrumentation is much more sensitive than the equipment available in the mid-20th century.

The effectiveness of these male pheromones was demonstrated using synthetic compounds in experiments designed to model courtship behaviour in the laboratory. In *P. napi* experiments male models, scented with or without artificial citral,

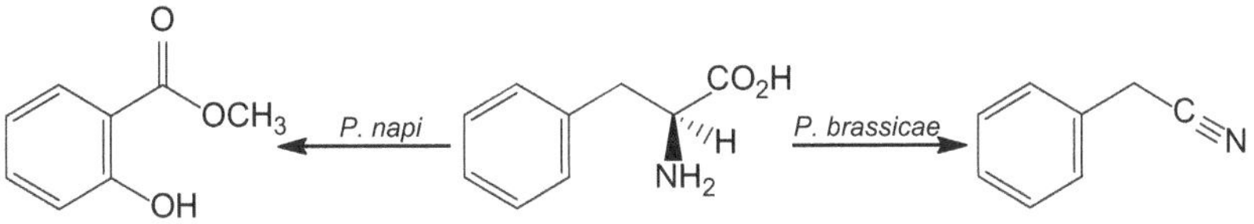

FIGURE 5.4 Biosynthetic origins of the anti-aphrodisiac compounds methyl salicylate and benzyl cyanide in two pierid butterflies.

were waved in front of female insects to simulate natural courtship whereas a mating-competition bioassay approach was used in *P. brassicae* studies in which live male insects were treated with either the pheromone mixture or a control substance.[2] Female insects were found to react in a more positive manner towards both types of pheromone-scented males compared to their untreated peers, either through a marked increase in mating success (*P. brassicae*) or by exhibiting higher levels of male acceptance behaviour (*P. napi*).[3] Moreover, both the geometric isomers geranial and neral components in citral were required for *P. napi* females to accept mating with a courting male. Whilst phytol and 6,10,14-trimethylpentadecan-2-one have been found in male scent organs of other Lepidoptera species, the third component, brassicalactone, is unique to the *P. brassicae* species.

During mating the *P. napi* and *P. brassicae* males transfer not only their sperm to the females but also the compounds methyl salicylate and benzyl cyanide respectively which are anti-aphrodisiacs (Figure 5.4). After mating the females release the anti-aphrodisiac compounds and they become cloaked in a volatile odour, thus turning them temporarily into a chemical mimic of the males and as a consequence they become unattractive to other males immediately after copulating. This unattractiveness has benefits to both the male and female; the male benefits by delaying female remating as long as possible to avoid sperm competition from other males whilst the female benefits as she has more time to invest in egg-laying without being courted by other males. The effect wears off over time and females can remate after four to six days. Male *P. napi* butterflies have a store of methyl salicylate which was synthesised from the amino acid L-phenylalanine (L-Phe) acquired during feeding in the caterpillar stage of their life. After their initial mating event the butterflies then replenish their stocks of the anti-aphrodisiac from volatile compounds in nectar which had been formed from the L-Phe precursor.[4]

The anti-aphrodisiac intra-species communication system indicates that mating has taken place but in the *P. brassicae* butterflies it is exploited by the parasitic wasp *Trichogramma brassica*. When this wasp detects the benzyl cyanide odour of a mated female butterfly, it hitches a ride to her egg-laying sites and then parasitises the freshly laid batch of eggs. Moreover, a closely related species *Trichogramma evanescens* also exploits the benzyl cyanide cue but only after learning the association between this chemical odour and the presence of food.

The aphrodisiac and anti-aphrodisiac pheromones of the pierid butterflies are secondary metabolites, compounds which are typical of a particular species or groups of organisms and which provide a competitive advantage to the individuals whilst not being essential for their fundamental life processes. This latter area is the realm of the primary metabolites, the sugars, amino acids, common fatty acids and nucleotides and the polymers derived from them (carbohydrates, proteins, lipids and nucleic acids) which are essential and ubiquitous in all organisms. Secondary metabolites are sometimes termed natural products and for centuries compounds extracted from plants and animals have been exploited by our species for their properties such as in medicines (e.g. morphine and penicillin), as colouring agents (e.g. alizarin and Tyrian purple) or as spices or perfumes. Whilst natural products have a vast diversity of molecular structures they are produced from a relatively small number of key building blocks (α-amino acids, acetate, mevalonate and intermediates of the shikimic acid pathway) which arise from the metabolic pathways for the primary metabolites (Figure 5.5).

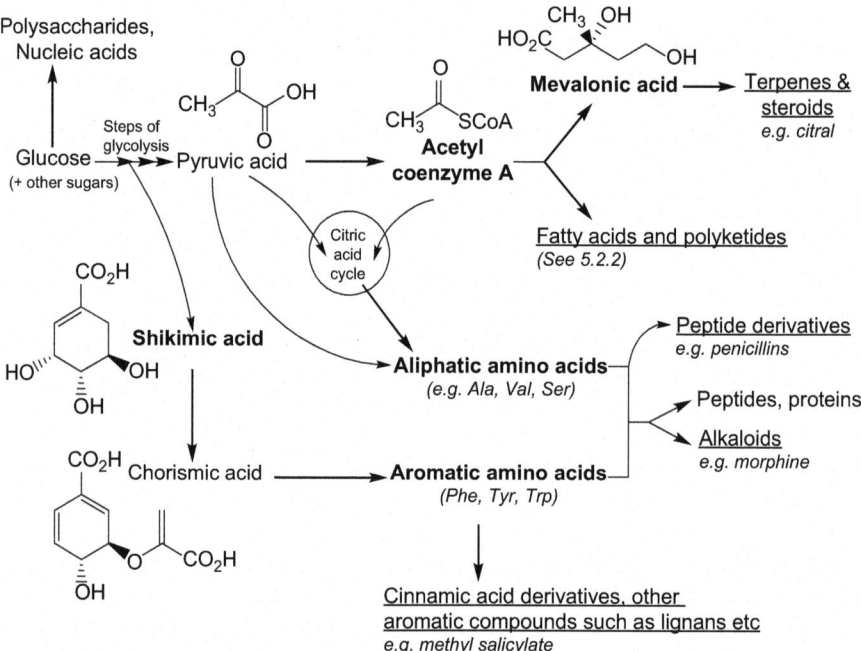

FIGURE 5.5 Outline of metabolic routes to primary and secondary metabolites. Key building blocks of secondary metabolites are shown in **bold** and secondary metabolites are underlined. At physiological pH the carboxylic acid group is ionised to the carboxylate ion, i.e. $RCO_2H \rightarrow RCO_2^-$. Individual steps in the biosynthetic pathways are catalysed by specific enzymes in collaboration with coenzymes which effectively function as reagents. For example, coenzyme A (CoA-SH) activates carboxylic groups as the thioesters, e.g. CH_3CO_2H is converted to acetyl coenzyme A, CH_3CO-SCoA. Some secondary metabolites are of mixed biosynthetic origin and are derived from building blocks from more than one of the main pathways. For example, shikimic acid and acetate building blocks lead to the flavonoids, a large class of plant polyphenolic compounds that includes anthocyanins which are responsible for the colour of red/blue fruit and vegetables.

5.2.2   A Blend of Three Compounds and a Moth Trap

The pine beauty moth, *P. flammea*, is a common species of nocturnal moth and the traditional food of the caterpillar is the needles of the Scots pine (*Pinus sylvestris*) tree. The very name *P. flammea* means destructive and flame-coloured and historically populations of this moth had not been a problem as its natural enemies kept the numbers under control. Commercial planting of the non-native lodgepole pine (*Pinus contorta*) species in the 1950s and 1960s, particularly the afforestation of peatland areas in Northern Scotland, provided the pine beauty moth caterpillars with an abundant alternative food source. This new habitat was an 'enemy-free space' for the caterpillars and they became a serious pest; widespread damage occurred to commercial plantations at regular intervals during the 1970s and 1980s which required intervention by spraying with insecticide. A decline in the frequency and impact of extensive tree defoliation in the lodgepole pine habitat since the 1990s has been largely attributed to an increase in natural predation and in particular a fungal pathogen which in some instances has caused over 88% mortality to the *P. flammea* caterpillars. Pheromone traps were introduced in 1993 to help monitor the populations of the adult moth during their flight period between late March and early May. A synthetic sex pheromone mixture is placed in a sticky trap to lure males which are then counted and a decision taken as to whether any further measures are necessary to keep the *P. flammea* population within the acceptable range.

Sexual communication in many moths is mediated by a precise mixture of fatty acid derivatives which vary by chain length, degree of unsaturation and the functional group at one end. The sex pheromone which the female *P. flammea* produces to attract males is a mixture of three compounds: (Z)-9-tetradecenyl acetate, (Z)-11-tetradecenyl acetate and (Z)-11-hexadecenyl acetate in a 100:1:5 ratio, i.e. compounds with 14 or 16 carbon atoms in the chain with one *cis* (or Z) double bond and an acetylated terminal alcohol group (Figure 5.6). Moreover, each of the three components has a distinctive role in the nocturnal quest of a male moth to first find, and then copulate with, a female partner. Experiments in the field and in the laboratory with synthetic pheromone molecules found that the main component (Z)-9-tetradecenyl acetate (Z-9:14Ac) plus either one or both of the two Z-11 compounds is the long-range attractant which allows the male to navigate towards the female. As the male gets closer to the female the concentration of the pheromone increases, but it is the presence of (Z)-11-tetradecenyl acetate (Z-11:14Ac) in the mixture which triggers the male to search for a landing site. After landing the male, guided by the Z-9:14Ac orientation cue, approaches the female and copulation is initiated by a high concentration of (Z)-11-hexadecenyl acetate (Z-11:16Ac) plus Z-9:14Ac. Copulation did not occur unless the Z-11:16Ac compound was present. Whilst all three pheromone components elicit the full flight-landing-copulation sequence, a binary combination of the main orientation compound, Z-9:14Ac, and the landing trigger, Z-11:14Ac, is sufficient as the bait to lure males into moth traps used to monitor *P. flammea* population levels in lodgepole pine plantations.

The even number of carbon atoms in the C_{14} and C_{16} compounds of the *P. flammea* pheromone reveals the acetate-derived origin of these molecules. Biosynthesis of the carbon chain involves a condensation reaction between two acetate-based units to form

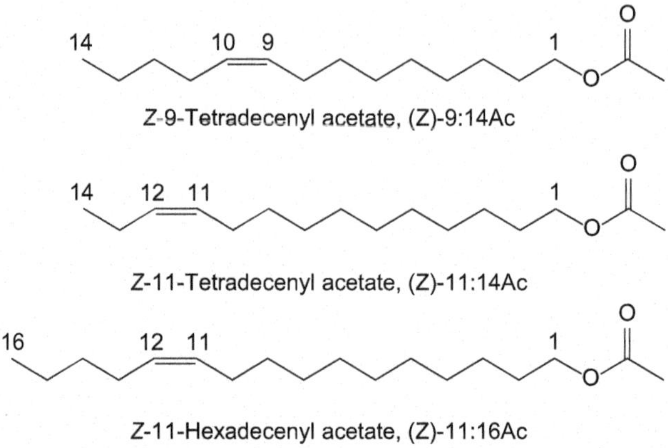

Z-9-Tetradecenyl acetate, (Z)-9:14Ac

Z-11-Tetradecenyl acetate, (Z)-11:14Ac

Z-11-Hexadecenyl acetate, (Z)-11:16Ac

FIGURE 5.6 Components of the *P. flammea* moth sex pheromone. Full names for the fatty acid molecules are given together with a shorthand version. The number before the colon indicates the position of the double bond from the functionalised carbon (assigned number 1) and most double bonds are of the Z (*cis*) configuration. The number after the colon indicates the hydrocarbon chain length and the group of letters specifies the functional group at the chain terminus. Thus, Acid = the carboxylic acid group (CO_2H), OH indicates a fatty alcohol, i.e. the chain terminates in CH_2OH, Ac indicates an acetate ester, i.e. O-$COCH_3$ at the end and Ald indicates a terminal aldehyde group.

a keto ester, CH_3–CO–CH_2–CO–SEnz, followed by a reduction process to convert the keto (CO) to a methylene (CH_2) group.[5] Repeated cycles of condensation and reduction thus assemble the chain two carbon atoms at a time until the required length is obtained. The completed chain is then cleaved from the enzyme to afford the free fatty acid, CH_3–$(CH_2$–$CH_2)_n$–CO_2H, that can undergo further modification to introduce double bonds into the molecule and reduce the carboxylic acid to the alcohol which can be esterified.

The bait used to lure male *P. flammea* moths into the monitoring traps comprises the two C_{14} compounds which differ only in the position of the Z-double bond, i.e. they have the structural formula of CH_3–$(CH_2)_x$–CH=CH–$(CH_2)_y$–OAc (*x*, *y* = integers). A Z-double bond is an important aspect of these compounds and there are a variety of methods available to the synthetic chemist to generate alkenes with this stereochemistry. Traditionally Wittig reactions have been used to construct Z-double bonds in which the alkene is put together from two halves, one in the form of the phosphorane RCH=P($C_6H_5)_3$ and the other in the form of an aldehyde O=CHR$_1$. The reaction is not completely stereospecific and the desired Z-product, RCH=CHR$_1$, can be contaminated with some E-isomer. Partial reduction of alkynes (RC≡CR$_1$) is another strategy and the reaction can be stopped at the Z-alkene stage by use of the Lindlar catalyst, a palladium catalyst deactivated by lead salts, to avoid unwanted over-reduction to the corresponding alkane.

Reduction of a triple bond to a Z-double bond can also be effected by a catalytic transfer semi-hydrogenation reaction with a hydrogen donor molecule, i.e. a hydrogen source other than molecular hydrogen. This reaction has the advantage that not only does it give good yields and stereoselectivity, but it avoids the production of triphenylphosphine oxide (a by-product of the Wittig reaction) or the use of toxic lead salts (Lindlar catalyst). The catalytic semi-hydrogenation reaction is often now used in the synthesis of the long-chain Z-alkene moth pheromones. Moreover, the Z-selective semi-hydrogenation reaction is an example of green chemistry as it is a more environmentally benign synthetic procedure than the more classical Wittig and Lindlar routes.

An efficient method to form the internal triple bonds required for the reduction reaction to the Z-alkene is the alkylation of a terminal alkyne ($R-C\equiv CH$). The hydrogen atom at the end of a terminal alkyne is acidic and can be removed by strong base to give the very reactive carbanion $R-C\equiv C^-$ which can displace a bromide ion from a halogenoalkane. This coupling reaction followed by Z-selective semi-hydrogenation of the internal alkyne are the two key steps in the synthesis of the target C_{14} compounds (Figure 5.7). A synthetic pheromone blend, typically containing 25 mg of Z-9:14Ac (the main orientation compound) and 2.5 mg of Z-11:14Ac (the landing trigger), was employed in the *P. flammea* sticky traps used in the monitoring of the adult male moths in lodgepole pine forests.

5.3 LIQUID CRYSTALS

5.3.1 A New State of Matter

Liquid crystals are substances with unusual melting properties. When these compounds are heated they do not pass directly from the solid phase to the liquid phase but rather go through an intermediate phase, sometimes referred to as the mesophase, which has some of the structure of solids but some of the freedom of motion of liquids (Figure 5.8). The process is reversible by lowering the temperature. Cholesteryl benzoate (Figure 5.9a) was the first compound reported to demonstrate this property in 1888; it melts at 145°C to form an opaque viscous liquid, the liquid crystalline state, and then changes to a transparent liquid at 179°C. These early investigations also found that physical properties such as the refractive index and electrical conductivity of the opaque mesophases depended on the direction in which they are measured, i.e. liquid crystals are anisotropic materials. As the liquid crystalline state has anisotropic properties characteristic of crystals, order in one or more spatial directions, and fluidity which is a property associated with liquids, these distinct mesophases are sometimes called a new state of matter (following gases, liquids and solids). Visual observation of the new state of matter of cholesteryl benzoate is one aspect of the experiment in Chapter 6 on this molecule.

Molecules that exhibit liquid crystalline behaviour are often long, thin and fairly rigid for some portion of their length. Intermolecular interactions enable the molecules to maintain an elongated shape that favours alignment and some degree of order in the mesophase. Early synthetic liquid crystals contained two rings linked together by a rigid linking group plus flexible hydrocarbon chains at each end, one of which contained a permanent electrical dipole (Figure 5.9b). Such liquid crystal molecules with a permanent dipole tend to align parallel to an applied electric field and it is the combination of the inherent anisotropic electrical and optical properties of liquid crystals which has led to their widespread application in screens for various electronic devices. Commercially successful liquid crystals commonly incorporate

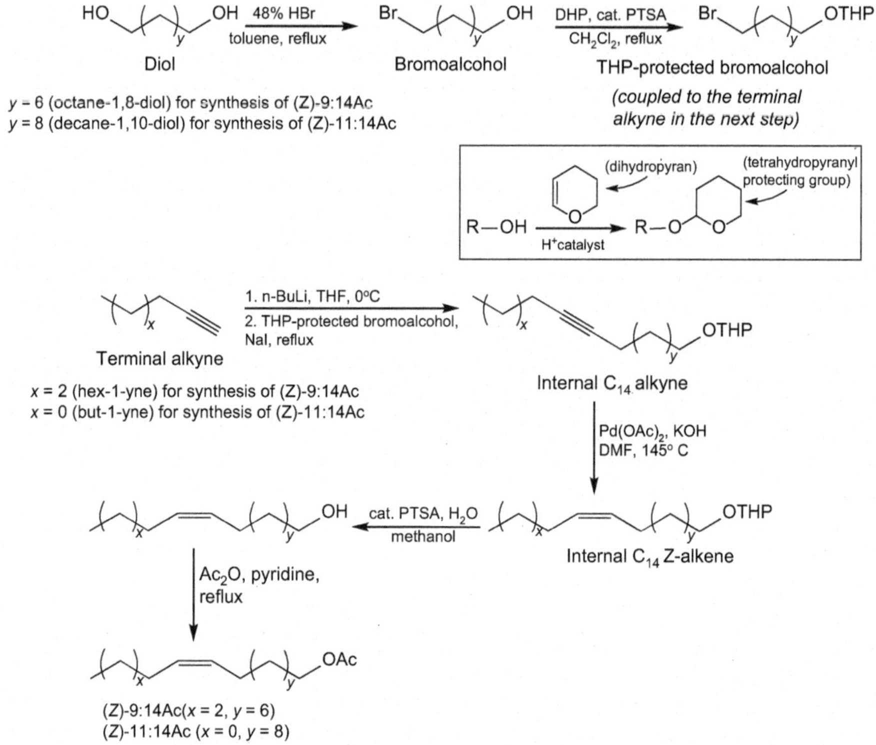

FIGURE 5.7 Chemical synthesis of long-chain aliphatic internal *Z* alkenes. The synthesis starts with monobromination of symmetrical diols with 48% HBr in refluxing toluene to afford the monobrominated alcohols. The hydroxyl groups of these compounds were then protected as the tetrahydropyran (THP) ether through reaction with dihydropyran (DHP) in the presence of an organic acid catalyst (*para*-toluenesulphonic acid, PTSA) (see inset). Protecting a hydroxyl group is desirable to avoid complications that can arise in subsequent reactions involving base. The coupling step in which two smaller molecules are joined to give the C_{14} compound with an internal alkyne group involved first treating either but-1-yne or hex-1-yne with *n*-butyllithium (a strong base) at 0°C in anhydrous tetrahydrofuran (THF) to give the corresponding negatively charged carbanion. This intermediate was alkylated with the protected bromoalcohol to afford the C_{14} alkyne through an S_N2 reaction. The addition of solid sodium iodide to the reaction mixture aids the S_N2 reaction as iodide is a good nucleophile and replaces the bromo substituent of the protected bromoalcohol in the reaction mixture and then also serves as a good leaving group for the alkylation. The crucial Z-selective semi-hydrogenation step involves DMF (*N,N*-dimethylformamide)/KOH as the hydrogen source system which forms formic acid (HCOOH) in situ by hydrolysis of the DMF [H–CO–N(CH$_3$)$_2$] solvent. The final steps involve deprotection of the alcohol using a catalytic amount of acid followed by esterification with acetic anhydride to give the final product.

two or three aromatic rings directly linked together with a non-polar moiety at one end of the molecule and a polar group at the other (Figure 5.9c).

Liquid crystals are used in display screens (see Section 5.3.2) because of their anisotropic properties such as birefringence. Birefringence or double refraction is an

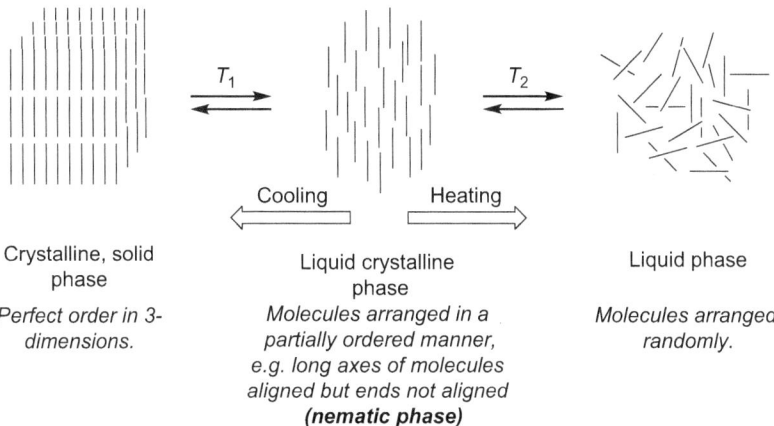

FIGURE 5.8 Molecular order in thermotropic liquid crystals.[10] When liquid crystals are heated and reach temperature T_1 the molecules adopt long-axis alignment with a one-dimensional degree of order. Some liquid crystals form a smectic phase at temperature lower than T_1 in which the aligned molecules are arranged in distinct layers giving order in two dimensions. Heating to T_2 transforms the nematic phase into the totally disordered isotropic liquid state.

optical anisotropy property which affects the way light travels through the material. As a ray of incident light strikes the surface of a birefringent material it is split into two refracted rays which propagate at different velocities through the material, i.e. there are two different refractive indices. Many minerals are also birefringent; in calcite the phenomenon is very strong with the two rays being separated when they emerge from the crystal. When a transparent calcite crystal is placed over an object the image observed is doubled (Figure 5.10).

The birefringent effect shown by calcite is a result of the internal arrangement of the ions in the crystalline lattice. Calcite ($CaCO_3$) has alternating layers of Ca^{2+} and trigonal planar CO_3^{2-} ions along the (long) c crystallographic axis (Figure 5.11b). As this sequence is distinct from other directions the calcite crystalline lattice is said to be anisotropic. In comparison, sodium chloride (NaCl) is optically isotropic as it has a cubic lattice structure in which each Na^+ ion is surrounded by six Cl^- ions and each Cl^- ion is surrounded by six Na^+ ions along three mutually perpendicular axes, i.e. the arrangement of ions is the same in all directions (Figure 5.11a).

The birefringence phenomenon of minerals can be observed in a polarising microscope which contains two polarising filters at 90° to each other on either side of the sample (Figure 5.11c and d). Ordinary light consists of transverse waves in which the vibrations are at right angles to the direction of travel. A polarising filter only permits vibrations oriented in a specific direction to pass through. When polarised light enters NaCl crystals the electromagnetic waves interact with an internal structure which is the same along all the crystallographic axes and thus passes through the structure at a single velocity, i.e. sodium chloride crystals have a single refractive index. The polarised light which has passed through the NaCl crystal is still polarised but is at

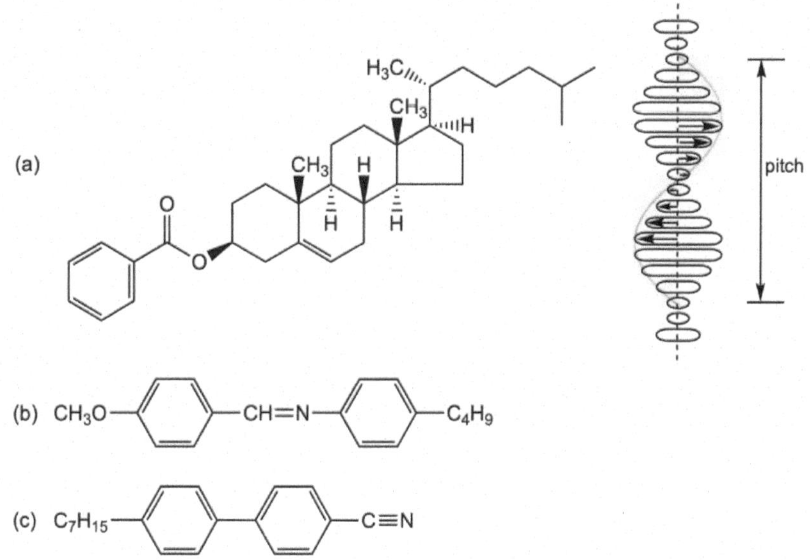

FIGURE 5.9   Molecular structures of liquid crystals. (a) Cholesteryl benzoate, an example of a cholesteric liquid crystal. Cholesteric liquid crystal molecules have a plate-like shape with chiral centres and align into a helical structure in which molecules are at a slight angle to those above and below it (indicated by the arrows). The distance along the helical axis within which the molecules complete a 360° rotation is known as the pitch and is typically in the range 0.1–10 μm which is much larger than the nanometre scale of molecules. The pitch of the helical structure is dependent upon temperature, becoming smaller as temperature increases which causes reflected light to contain shorter wavelength components of the incident light. When the pitch of this helical arrangement is comparable to the wavelength of visible light a cholesteric liquid crystal may reflect light (Bragg reflection) generating striking colours. The thermochromic effect of mixtures of cholesterol derivatives is used in liquid crystal thermometers. (b) N-(4-methoxybenzylidene)-4-butylaniline (MBBA, mesophase 20°C–47°C), a typical example of an early synthetic liquid crystal. This compound is chemically not very stable as it easily hydrolyses. (c) 4-Cyano-4′-pentylbiphenyl (5CB, mesophase 22°C–35°C), one of a series of cyanobiphenyl compounds which possess a stable nematic phase at room temperature and made LCDs viable.

right angles to the permitted vibration direction of the second polariser and thus no light is transmitted to the observer.

Polarised light passing through a crystal of calcite along the c crystallographic axis behaves in a manner similar to the interaction occurring with isotropic crystals, i.e. the birefringence has no effect. However, when light enters the calcite crystal in a direction perpendicular to the c axis the anisotropic lattice splits the polarised light into two rays at right angles to each other. Each ray encounters a different atomic arrangement and therefore travels at a different speed through the crystal. The light which emerges from the crystal vibrates in planes that are mutually perpendicular and some light is transmitted through the second polarising filter and the calcite is visible to the observer.

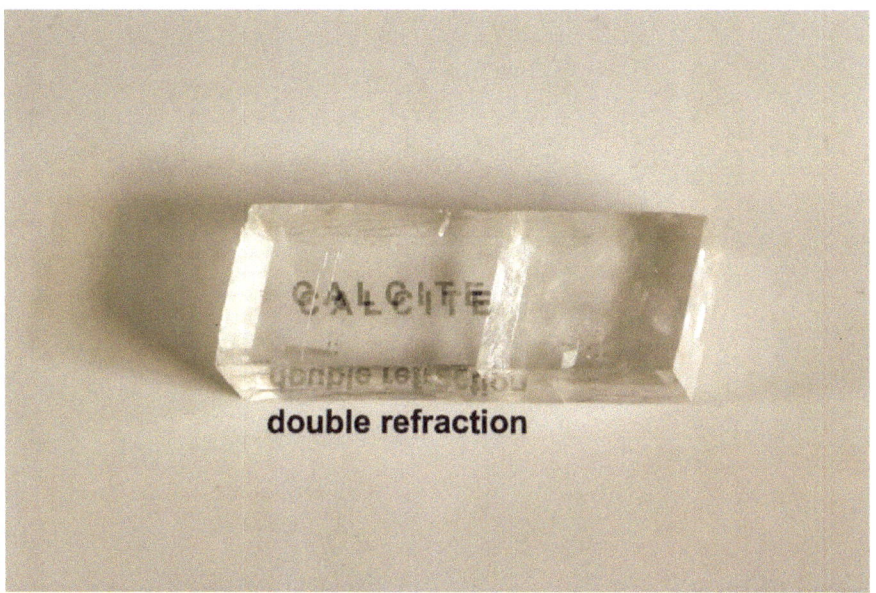

double refraction

FIGURE 5.10 The birefringence effect of calcite.

5.3.2 Liquid Crystal Displays

Liquid crystals remained largely a scientific curiosity and a niche area of research until the mid-20th century. By the 1960s there was a growing need to develop light-weight, power-efficient screens as an alternative to high-voltage bulky cathode ray displays used at the time and this objective, together with availability of new electronic devices such as the transistor and integrated circuit, revitalised research into liquid crystals. Pivotal breakthroughs included the discovery of the twisted nematic mode and the synthesis of the cyanobiphenyl series of stable liquid crystals which exhibited a low melting nematic phase. These inventions paved the way for the first pocket calculators and digital watches with liquid crystal displays (LCDs) in the 1970s.

In the early years of LCDs a family of organic compounds known as Schiff bases, exemplified by N-(4-methoxybenzylidene)-4-butylaniline (MBBA; Figure 5.9b), was extensively explored; these were birefringent materials, exhibited liquid crystallinity within the required temperature range and possessed a dipole moment. However, the central –C=N– link between the two aromatic rings was easily hydrolysed and the compounds were not sufficiently stable for commercial LCDs. Eliminating this group by directly linking the two phenyl rings and incorporating a cyano (–C≡N) group into one of the rings gave the chemically stable cyanobiphenyl family of compounds. The presence of the cyano group gave these compounds a strong longitudinal permanent dipole and this dielectric anisotropy was crucial for a twisted nematic device (see below). No single liquid crystal compound meets the temperature range required for an LCD and eutectic mixtures are used in these devices. A eutectic mixture is a

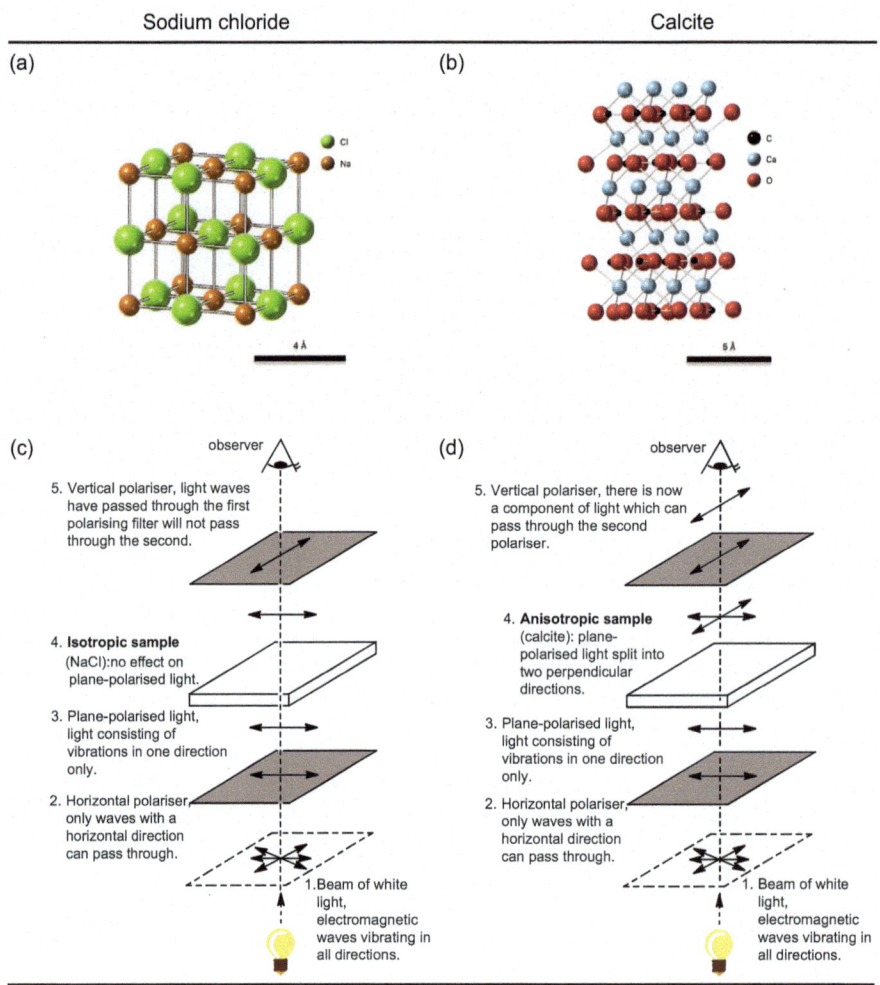

FIGURE 5.11 Minerals and birefringence. Crystal structures of (a) NaCl and (b) calcite, CaCO₃. Birefringence of (c) isotropic and (d) anisotropic samples between crossed polarisers. Geological samples such as calcite for viewing in a polarising microscope are a standard thickness of 30 μm and the minerals within a rock sample will generally have a random orientation. Furthermore, in a polarising microscope the sample stage and/or the second analyser can be rotated. In rocks containing anisotropic minerals the two orthogonal components of light travelling through the specimen at different speeds emerge out of step and this difference causes a phase shift; some wavelengths of light may be in phase whilst others may be out of phase. The second polariser recombines components of the two rays travelling in the same direction and vibrating in the same plane. Constructive and destructive interference of light occurs which produces colour in the mineral grains. The interference colours, and the change in their intensity which occurs when the specimen is rotated, are useful diagnostic features to aid the identification of mineral and rock samples. The crystal structures of sodium chloride (a) and calcite (b) were produced using CrystalViewer® 11.2.2. (CrystalMaker Software Limited, Begbroke Science Park, Oxfordshire, UK.)

combination of compounds that form a solid phase with a melting point much lower than that of any of the individual components. The optimum cyanobiphenyl eutectic system was E7: a mixture of four compounds consisting of three biphenyls and one terphenyl which met the stability and operating temperature range (−10°C to +60°C) required by display manufacturers (Table 5.1).

When plane polarised light passes through the nematic phase of a liquid crystal it changes the direction of polarisation by a certain angle. The thickness of the liquid crystalline layer, typically around 8–10 μm, can be controlled to produce a rotation of polarisation of exactly 90°. Therefore, when a cell containing the liquid crystalline mixture is placed between crossed polarisers this arrangement allows light to pass through; this is the basis of the twisted nematic LCD that was invented in 1971.

In the twisted nematic cell the liquid crystals are sandwiched between two orthogonally micro-grooved plates (Figure 5.12). The lozenge-shaped molecules naturally orient themselves in line with a chemical surfactant spread onto the grooves of each plate which are at 90° to each other and this forces a 90° helical twist across the nematic liquid crystal medium. The cell lies between a parallel pair of crossed polarising filters which are coated on the outside with a film of transparent conducting material, often indium tin oxide, which forms the electrodes. Plane polarised light entering

---

TABLE 5.1
Composition of the E7 Eutectic Liquid Crystalline Mixture

Compound	Nematic Phase Temperature Range (°C) of Pure Compound	Amount (%)
C_5H_{11}—⬡—⬡—$C\equiv N$	22–35	51
C_7H_{15}—⬡—⬡—$C\equiv N$	28–42	25
$C_8H_{17}O$—⬡—⬡—$C\equiv N$	54–80	16
C_5H_{11}—⬡—⬡—⬡—$C\equiv N$	130–239	8

These compounds were designed and developed in the early 1970s by the team at Hull University led by George Gray and E7 remained a preferred mixture for LCDs for many years. Incorporation of a terphenyl component into the mixture increased the transition temperature from the nematic phase to an isotropic liquid, giving a wider range liquid crystalline material than was attainable using biphenyl compounds alone.

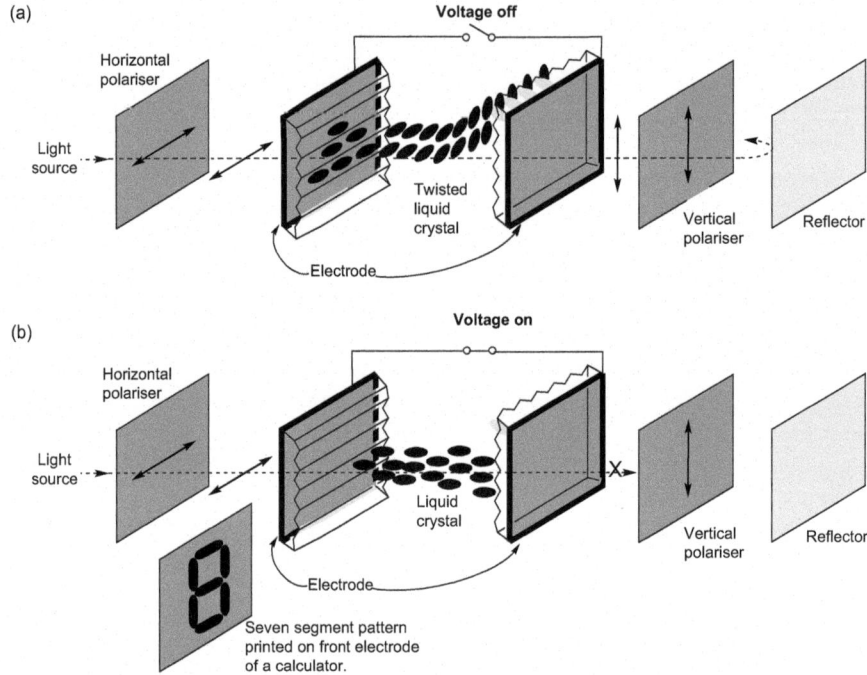

FIGURE 5.12 Liquid crystals as on-off light value switches. (a) Voltage-off state: external light shines onto a horizontal polariser, the polarisation is redirected by 90° by the nematic molecules so light passes through the vertical polariser. The light, reflected from the reflector at the rear, tracks back its path in reverse and the observer sees a bright region of light. (b) Voltage-on state: when a voltage is applied, the nematic molecules straighten out of their twisted orientation and no longer redirect the polarisation. Light remains polarised in its original direction and is unable to pass through the second polariser.

the liquid crystals is rotated (birefringence effect) and the helical twist guides the light through the layers of molecules such that it is in the correct orientation to pass through the second polarising filter. This is the voltage-off state and the display is transparent. When a voltage is applied to the plates, the liquid crystalline molecules align with the electrical field (dielectric effect) that directly runs between the electrodes thus unwinding the twist, the birefringence is lost and light will not be able to pass through the second polariser. This is the voltage-on state and the display becomes black. When the field is turned off, the crystals will relax back into the twisted structure and light will again be able to pass through.

A pattern of seven separate electrodes is printed on the front electrode of a calculator to generate the numerals 0 to 9. When a voltage is applied between any one or more of these segments and the back plate, the liquid crystal molecules under only those segments will be aligned in a straight line rather than twisted and the areas appear black. The number 2, for example, is generated by switching off segments at the top left and bottom right.

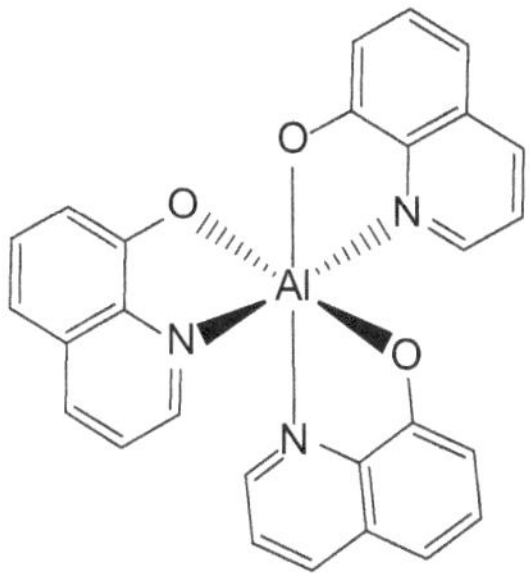

Tris(8-hydroxyquinoline) aluminium(III)

FIGURE 5.13   Structure of tris(8-hydroxyquinoline) aluminium(III) (Alq$_3$), a metal chelate complex of 8-hydroxyquinoline in which one aluminium atom is bound to three 8-hydroxy-quinoline ligands in a bidentate manner. The nitrogen and oxygen atoms of the 8-hydroxy-quinoline ligand are coordinated to the aluminium ion so that the complex takes on an octahedral configuration.

Many modern LCDs use a light source, often a white light-emitting diode in place of the reflector, as the source of illumination in order to obtain brighter images but the principle is the same as shown in Figure 5.12 except that the light moves from the back towards the viewer at the front. Thin-film transistor (TFT) matrix technology enables voltages to be applied to very small areas of the screen compared to the seven-segment system of calculator displays, thus vastly increasing the amount of the information content which can be displayed. The screen is divided into a large number of tiny squares (pixels) and each pixel is individually controlled (on/off) by one TFT. The amount of charge needed to control each TFT is very small which allows the entire display to be refreshed very quickly, and so can change rapidly with respect to the response times of the human eye. Colour filters are used to create red, green and blue subpixels and by varying the amount of light which passes through each subpixel a wide range of colours can be generated.

LCDs superseded cathode ray displays, and it is likely that LCDs will in turn coexist with and then be replaced by other emerging technology. One current candidate is the organic light-emitting diode (OLED) display in which an organic compound emits light in response to an electric current. As this system does not need a backlight the screens can be thinner and lighter than those based on liquid crystals. Tris(8-hydroxyquinoline) aluminium(III) (Alq$_3$, Figure 5.13) is one of the most successful organic materials used in OLEDs and the development of such devices relies on close collaboration between chemists, physicists and technologists. A synthesis of Alq$_3$ is one of the experiments in Chapter 6.

5.4   BATTERIES

Commercial batteries are self-contained electrochemical units that are used as a source of electrical power. Oxidation-reduction (or redox) reactions within the battery unit generate a flow of electrons which is harnessed to produce the current which powers devices. Batteries come in a myriad of shapes and sizes with voltages and performance characteristics designed to meet specific needs. For example, the 12 V lead-acid car battery is capable of delivering a large initial current to start an internal combustion engine whilst the cylindrical 1.5 V alkaline Zn-MnO$_2$ (AA) batteries supply small amounts of current over a longer period of time and are widely used in many household appliances.[6] Some batteries such as the alkaline Zn-MnO$_2$ batteries are thrown away when they can no longer produce a useable current, whilst others, the lead-acid car battery for example, are rechargeable. Batteries are (usually portable) packages containing voltaic cells. In an individual voltaic cell the transfer of electrons that occurs in a redox reaction takes place through an external circuit rather than directly between reactants present in the same reaction vessel. The movement of electrons in this external circuit is the current that is used to power electrical devices. All voltaic cells work in the same general way and the basic principles by which they produce an electric current are described in Section 5.4.1. The redox chemistry of some common single-use and rechargeable batteries is then outlined with lithium-ion batteries being considered separately in Section 5.4.2.

5.4.1 Voltaic Cells: Principles and Practice

In any redox reaction both oxidation and reduction must occur; if one substance is oxidised another must be reduced. For example, when a piece of zinc metal is placed into an aqueous solution of copper sulphate, the blue colour characteristic of aqueous Cu^{2+} starts to fade, the solution becomes hot and the surface of the zinc acquires a dull reddish colour as metallic copper is deposited. At the same time the zinc begins to dissolve and these transformations are summarised in the equation Zn (s) + Cu^{2+} (aq) → Zn^{2+} (aq) + Cu (s). The thermal energy generated in this reaction dissipates and is wasted. Although the oxidation and reduction reactions must take place simultaneously, it is often convenient to consider them as separate processes, i.e. zinc metal is oxidised and copper ions are reduced:

$$\text{Oxidation}: \text{Zn}(\text{s}) \rightarrow \text{Zn}^{2+}(\text{aq}) + 2\text{e}^- \left[\text{or Zn}(\text{s}) - 2\text{e}^- \rightarrow \text{Zn}^{2+}(\text{aq}) \right]$$

$$\text{Reduction}: \text{Cu}^{2+}(\text{aq}) + 2\text{e}^- \rightarrow \text{Cu}(\text{s})$$

When the reaction occurs in a voltaic cell apparatus (Figure 5.14) the components of the oxidation and reduction reactions are kept in separate compartments called half-cells. Thus the zinc half-cell on the left contains zinc metal that serves as one electrode and a solution containing Zn^{2+} (aq) ions and the half-cell on the right uses a copper electrode and a solution containing Cu^{2+} (aq) ions. The electrodes are joined by a conducting material such as copper wire (called the external circuit) through which the electrons are transferred from the electrode where the oxidation occurs to the one where reduction occurs. The two half-cells are connected by a salt bridge (the internal circuit) to keep the solutions electrically neutral and complete the circuit. When the concentration of Zn^{2+} (aq) and Cu^{2+} (aq) are 1.0 mol/dm^3 and the cell

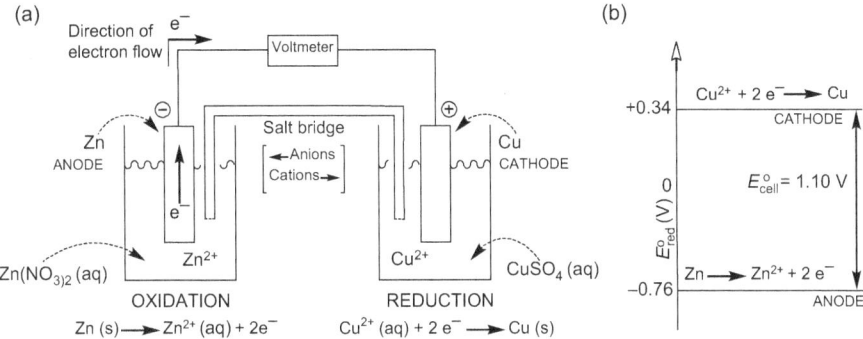

FIGURE 5.14   A simple voltaic cell based on the reaction between zinc metal and aqueous copper(II) ions. (a) One half-cell contains zinc metal dipping into a solution containing Zn^{2+} (aq) ions and the other has copper metal dipping into a solution of Cu^{2+} (aq) ions. The zinc metal electrode provides a path for the electrons produced in the oxidation reaction to be transferred to the copper electrode where they reduce Cu^{2+} (aq) ions in solution near the surface of the electrode. The electrons produced in the oxidation reaction give the anode a negative charge, indicated by the minus sign, and the cathode has a positive charge, or a plus sign. The salt bridge contains an electrolyte such as aqueous KNO_3 solution which allows anions and cations to move between the two cells so that the number of positive and negative charges in each half-cell remains balanced. As electrons move from anode to cathode in the external circuit the negative anions move through the salt bridge from the cathode to the anode compartment and the cations move in the reverse direction. Thus negative NO_3^- ions migrate towards the anode and positive K^+ ions migrate towards the cathode. (b) Each half-cell is assigned a standard reduction potential value which provides a measure of the tendency for the reduction reaction to occur and the more positive the value of E_{red}^{o}, the greater the tendency for the reduction reaction. E_{red}^{o} values cannot be measured separately and so are determined with respect to a standard; this is the hydrogen electrode which is assigned a value of zero.

is at 25°C a cell potential of 1.10 V is generated. In battery-operated devices the voltmeter is replaced by the device that operates by using electrical energy.

As the redox reaction proceeds, the reactant concentration decreases and the product concentration increases which, when carried out in a voltaic cell, causes the voltage produced to vary. Simple voltaic cells such as those described above are neither compact nor robust and are not useful in the real world. Batteries are commercial versions of simple voltaic cells and overcome their limitations. There are two broad categories of batteries: secondary cells in which the redox reactions can be reversed, i.e. the battery can be recharged, and primary cells where the redox reactions cannot be reversed and this type is discarded when they are discharged.

The cylindrical AA or button-shaped LR44 alkaline Zn-MnO_2 batteries fall into the latter category. As current is drawn from the battery (discharging) the MnO_2 incorporated into the graphite cathode is reduced, the Zn anode is oxidised and the Zn^{2+} ions react with the basic (usually concentrated KOH) electrolyte to form $Zn(OH)_2$. The cell reactions are complex and can be approximated to those shown in Figure 5.15a, but they are not reversible.

(a) <u>Alkaline Zn-MnO$_2$ cell</u> *(single use)*

Oxidation *(anode)*: $Zn (s) + 2 OH^- (aq) \longrightarrow Zn(OH)_2 (s) + 2 e^-$

Reduction *(cathode)*: $2 MnO_2 (s) + 2 H_2O (l) + 2 e^-$
$$\longrightarrow 2 MnO(OH) (s) + 2 OH^- (aq)$$

Overall reaction: $Zn (s) + 2 MnO_2 (s) + 2 H_2O (l)$
$$\longrightarrow Zn(OH)_2 (s) + 2 MnO(OH) (s)$$

(b) <u>Lead-acid</u> *(rechargeable)*

Oxidation *(anode)*: $Pb (s) + SO_4{}^{2-} (aq) \xrightleftharpoons[\text{charge}]{\text{discharge}} PbSO_4 (s) + 2 e^-$

Reduction *(cathode)*: $PbO_2 (s) + 4 H^+ (aq) + SO_4{}^{2-} (aq) + 2 e^- \xrightleftharpoons[\text{charge}]{\text{discharge}} PbSO_4 (s) + 2 H_2O (l)$

Overall reaction: $Pb (s) + PbO_2 (s) + 2 H_2SO_4 (aq) \xrightleftharpoons[\text{charge}]{\text{discharge}} 2 PbSO_4 (s) + 2 H_2O (l)$

(c) <u>Nickel-metal hydride</u> *(rechargeable)*

Oxidation *(anode)*: $MH (s) + OH^- (aq) \xrightleftharpoons[\text{charge}]{\text{discharge}} M (s) + H_2O (l) + e^-$ (M = metal alloy)

Reduction *(cathode)*: $NiO(OH) (s) + H_2O (l) + e^- \xrightleftharpoons[\text{charge}]{\text{discharge}} Ni(OH)_2 (s) + OH^- (aq)$

Overall reaction: $NiO(OH) (s) + MH (s) \xrightleftharpoons[\text{charge}]{\text{discharge}} Ni(OH)_2 (s) + M (s)$

FIGURE 5.15   Cell reactions for some common batteries. The oxidation/reduction labels in the reactions refer to the discharge reactions. Alkaline Zn-MnO$_2$ cells are not rechargeable as the products from the discharge reaction disperse into the cell electrolyte and cannot be 'unmixed' for the reverse reaction.

Lead-acid batteries are widely used in cars with internal combustion engines and are the oldest type of rechargeable battery, being first commercialised in the late 19th century. The electrodes are lead (anode) and lead(IV) oxide, PbO$_2$ (cathode), which are immersed in sulphuric acid (~30%) as the electrolyte. The product from both the oxidation and reduction reactions is (solid) lead(II) sulphate, PbSO$_4$ (Figure 5.15b) which adheres to the electrode and it is this feature which makes this battery rechargeable. During recharging an external source of electrical energy drives the nonspontaneous reverse reaction thus regenerating Pb and PbO$_2$. Recharging is usually performed by the vehicle's alternator when there is no current demand on the battery. The electrolysis equation for recharging a voltaic cell is the reverse of that for the discharge process. Eventually the amorphous form of PbSO$_4$ becomes hard and crystalline and cannot be converted back into Pb and PbO$_2$, so the battery has to be replaced.

Development of the nickel-metal hydride (Ni-MH) battery in the late 20th century produced a rechargeable 1.25 V cell which can often be used as a substitute for similarly shaped non-rechargeable alkaline batteries. The Ni-MH battery uses a metal alloy as the anode, typically a lanthanum compound such as LaNi$_5$, which can absorb hydrogen atoms into interstitial (unoccupied) positions in the metal lattice. During oxidation, the hydrogen atoms are desorbed, lose electrons and the resultant

H^+ ions react with OH^- ions of the electrolyte (30% KOH) to form H_2O. At the positive electrode, nickel(III) oxyhydroxide, NiO(OH), is reduced to nickel(II) hydroxide, $Ni(OH)_2$, as shown in Figure 5.15c. Ni-MH batteries are used in some hybrid electric vehicles and are recharged by the electric motor whilst braking.

5.4.2 Lithium-Ion Batteries: A Rocking Chair Design

The power demands of mobile phones and laptops would drain a nickel-metal hydride battery in less than an hour but the chemistry of the lithium-ion batteries enables many mobile electronic devices to function all day without recharging. Lithium-ion batteries have also been produced to power electric vehicles and are used as storage devices for intermittent renewable energy sources such as photovoltaic cells in solar panels. John B. Goodenough, M. Stanley Whittingham and Akira Yoshino were awarded the 2019 Nobel Prize in Chemistry in recognition of their seminal work developing the lithium-ion battery.

Lithium-ion batteries work by coupling electron-transfer redox reactions with the movement of lithium ions between a graphite anode and a lithium cobalt oxide ($LiCoO_2$) cathode (Figure 5.16). During battery charging using an external electricity

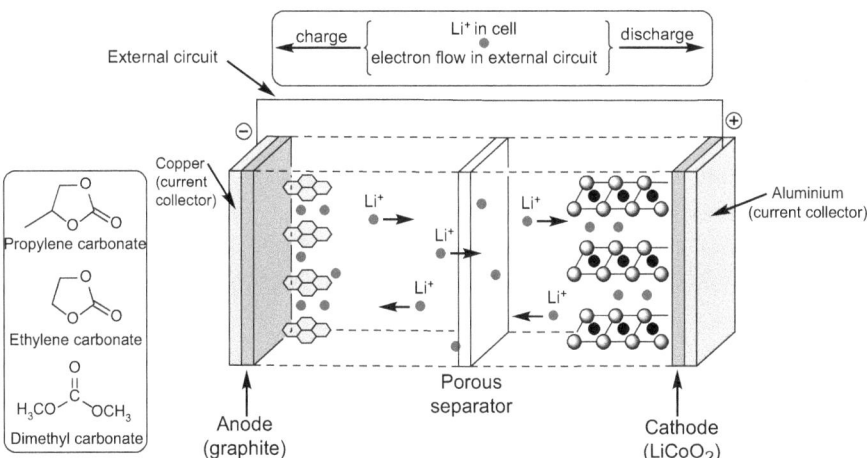

FIGURE 5.16 Components of a rechargeable lithium-ion battery. In Li-ion batteries the electrodes are porous composite materials and thin films of these are coated onto copper or aluminium foils to provide a physical support and to collect and deliver electrons between electrodes via the external circuit. The two electrodes are separated from each other by a porous separator and are immersed in a non-aqueous liquid electrolyte which is the conductive medium for the migration of the lithium ions. The electrolyte is a combination of a lithium salt, usually lithium hexafluorophosphate ($LiPF_6$) dissolved in one or more organic carbonates (see inset for compound structures). Note that the positive electrode is always referred to as the cathode and the negative electrode is always the anode regardless of whether the cell is charging or discharging. Interfacing the active components (anode, cathode, electrolyte) with the more inert materials (current collectors, separators, additives etc.) so that they all work well with each other is a challenging and integral part of battery development but beyond the scope of this chapter.

supply, some cobalt Co(III) ions in $LiCoO_2$ are oxidised to the cobalt(IV) species. To maintain the charge balance, some of the positively charged Li^+ ions leave the $LiCoO_2$ lattice structure, dissolve in the electrolyte solution and travel to the graphite side where they are inserted between the layers of carbon atoms. This reaction also deposits electrons into the graphite structure via the external circuit. During discharge when the battery is powering portable electronic devices the reverse process occurs; lithium ions move back to the metal oxide whilst electrons are released from the carbon electrode to the external circuit where they do useful work before entering the cathode and reducing Co(IV) to Co(III) ions, thus regenerating the original $LiCoO_2$ structure. These reactions can be summarised by the following equation and the first commercial lithium-ion battery was introduced by the Sony company in 1991.

$$LiCoO_2 + C_6 \text{ (graphite)} \underset{\text{discharge}}{\overset{\text{charge}}{\rightleftharpoons}} Li_{1-x}CoO_2 + Li_xC_6 \text{ (intercalated graphite)}$$

(5.1)

A reversible ion transport and insertion mechanism between separated redox active electrodes was the novel idea which initiated the development of rechargeable lithium batteries. Evolution of this concept, sometimes referred to as a 'rocking chair' mechanism, into a lightweight and high-energy-density power source entailed innovative solutions to many problems which arose during the development of the anode, cathode and electrolyte formulation. These problems and solutions are presented below as a series of snapshots where the focus is on the chemical nature of the anode, cathode and electrolyte and the relationships between the components.

5.4.2.1 Storing Lithium Ions: An Intercalation Cathode

Alkali metals readily lose electrons which makes them good candidates for the anode in voltaic cells and metallic lithium is used in some primary (non-recyclable) batteries, e.g. CR123A. In a lithium rechargeable cell the lithium ions produced by the oxidation at the anode need to be stored and made available at the cathode for conversion back to their atomic form in the charging (reduction) step. A solution to this storage issue came in the form of intercalation compounds. In these substances a guest species (atom, ion or molecule) is trapped within the lattice structure of a host compound, with the structure of the latter species remaining unchanged.

Titanium disulphide (TiS_2) was selected as a promising host substance to store intercalated lithium ions on the basis that it was a lighter weight counterpart to tantalum disulphide (TaS_2), a compound which had previously been found to intercalate potassium ions during a project focused on superconductivity. Titanium disulphide has a structure in which the sulphide ions are in a hexagonal close-packed array with the Ti^{4+} ions occupying octahedral holes, i.e. each Ti^{4+} ion is surrounded by six covalently bonded sulphide ligands (Figure 5.17a). Extending this lattice gives a 'stacked sandwich' structure; each sandwich consists of a layer of S^{2-} ions, a parallel layer of Ti^{4+} ions and another parallel layer of S^{2-} ions with weak van der Waals forces between the stacked sandwiches (Figure 5.17b). Lithium ions can intercalate into this central gap between the sandwiches to give Li_xTiS_2 ($x < 1$). Moreover, electrochemical studies demonstrated that this process was reversible; lithium ions could

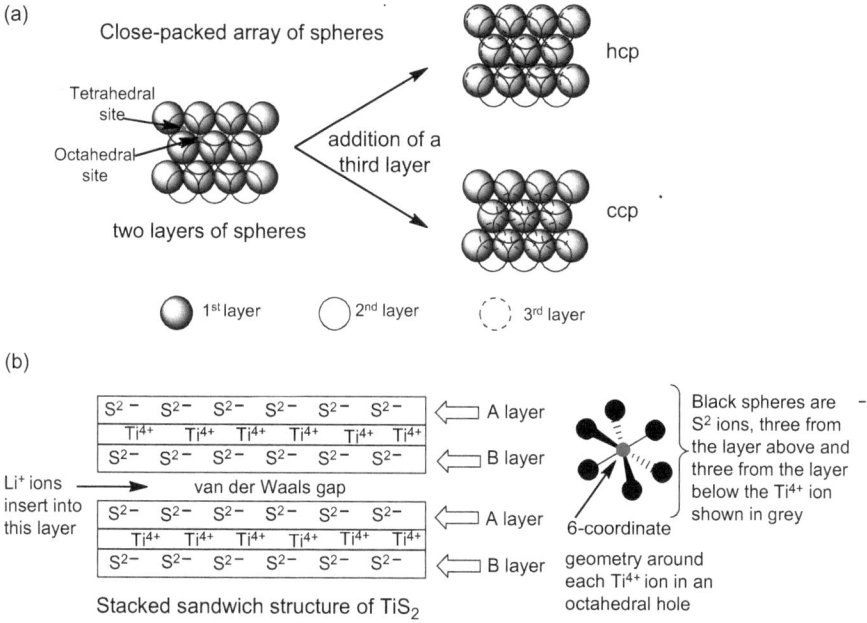

FIGURE 5.17 Solid state structures. (a) Close-packing of spheres. In layered lattices such as TiS_2 the largest ion (usually the anion) is close-packed. In each layer of a close-packed structure each sphere or ion has six neighbours in a hexagonal array. The spheres of a similar second layer fit into the depressions in the first layer as indicated in the diagram on the left. There are two possible ways of adding a third layer; directly over the first layer or in a third position as represented by the diagram on the bottom right. The former of these two arrangements gives an ABAB... sequence which is hexagonal close-packing (hcp) and the latter arrangement gives an ABCABC... sequence which is cubic close-packing (ccp). Smaller cations fit into some of the spaces between the layers: either in a tetrahedral site between four spheres or the larger octahedral sites between six of the spheres. (b) Schematic of two 'stacked sandwich' layers of the TiS_2 with a hcp ABAB sulphur lattice with Ti^{4+} ions sitting in octahedral holes in every other layer.

be inserted into and removed from the gap without significantly disrupting the overall crystal structure of TiS_2. The first rechargeable lithium cell constructed, but not commercialised, combined a metallic lithium anode, a TiS_2 cathode and an organic electrolyte ($LiPF_6$ dissolved in propylene carbonate) and had a voltage of about 2.4 V.

There are safety issues with using lithium metal as an electrode in this rechargeable battery as lithium dendrites, branching tree-like structures, can grow during battery charging leading to short-circuiting and fires. Further development work could not satisfactorily overcome the safety concerns and rechargeable lithium metal batteries were abandoned in favour of the design based on lithium ions. However, this new strategy presented the problem of not only developing an insertion-compound anode but also an alternative cathode. As a Li^+/Li electrode possesses the largest potential for an oxidation reaction, this alternative anode material with a smaller oxidation potential would therefore lower the output voltage to a value where the battery may not be competitive with the Ni-MH system.[7] Therefore, a new cathode material that could also intercalate lithium ions at a high reduction potential was required to generate a battery with an acceptable voltage.

5.4.2.2   An Intercalating Oxide Cathode

The switch from using a TiS_2 cathode to the oxide equivalent stemmed from Goodenough's insight that the top of the S^{2-} 3p valence band was at a higher energy than the top of the O^{2-} 2p band.[8] As cell voltage is determined by the difference between the redox energies of the anode and cathode, if the redox behaviour of the transition metal could be linked to a lower-lying energy band then the operating voltage could be increased.

However, transition metal oxides do not form layered structures as the repulsive forces between oxide-ion layers are larger than the binding energy from van der Waals forces and TiO_2, for example, can adopt a rutile crystal structure (Figure 5.18a). Recalling earlier work on an unrelated project on the structure of mixed metal oxides, the research team shifted their investigations to the $LiMO_2$ family of compounds, particularly those containing cobalt and nickel ions, which do have a layered structure (Figure 5.18b). $LiCoO_2$ was found to be the best intercalation cathode material as the

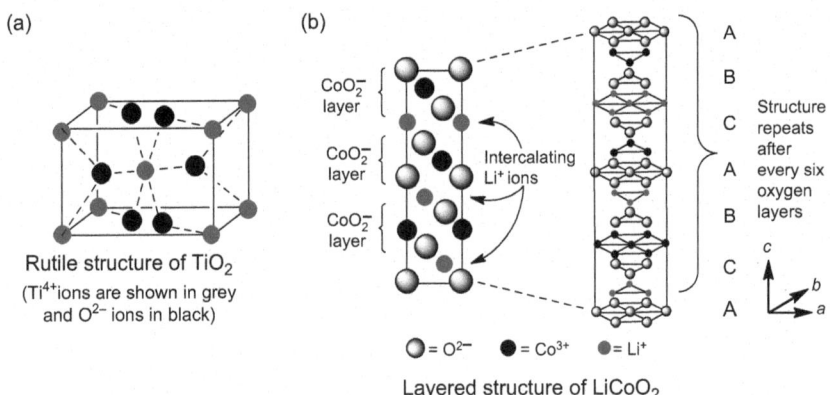

FIGURE 5.18   Metal oxide structures. (a) Rutile structure of TiO_2. The parallelopiped shown is the unit cell which is the three-dimensional repeat unit of the crystal lattice and TiO_2 does not form a close-packed structure. The coordination numbers of titanium and oxygen are 6 (octahedral) and 3 (trigonal planar) consistent with the 2:1 stoichiometry of TiO_2. The rutile structure type is also adopted by other metal oxides such as cassiterite, SnO_2. (b) Layered structure of $LiCoO_2$. The oxygen layers are organised in a cubic close-packed arrangement (ABCABC stacking sequence) in which each layer is displaced by a third from the previous layer. The Li^+ and Co^{3+} ions occupy interstitial octahedral sites in alternating layers in the ABC repeat unit of the oxygen framework. As the oxygen stacking has a repeat unit of three and the metal layering repeats every two layers, periodicity is achieved after six oxygen layers. In the three-dimensional structure this stacking arrangement of ions generates edge-sharing CoO_6 octahedra linked together in the ab-plane to form CoO_2^- layers which are screened and stabilised by the octahedrally coordinated intercalated Li^+ ions.

Li^+, Co^{3+} and oxide ions formed completely separate and ordered layers which was better suited to accommodate the changes in lithium-ion concentration during the de-intercalation process. The situation was more complicated in $LiNiO_2$. This compound is non-stoichiometric with some nickel ions always being found in lithium sites which disturbs the layered structure motif and hinders the diffusion pathway of the lithium ions. Furthermore, cation mixing also leads to structural instability during charge-discharge cycles with the formation of a non-layered spinel species $LiNi_2O_4$.

5.4.2.3   A Mismatch and a Compromise

A stable battery requires the holistic integration of anode, cathode and electrolyte. The electrolyte facilitates the transport of lithium ions and must be selected such that it does not react with either electrode. Decomposition of the electrolyte at the electrodes leads to decreased cell performance and lifetime. If the HOMO (highest occupied molecular orbital) energy of the electrolyte is above the electrode potential of the cathode (μ_C), electrolyte molecules lose electrons and are oxidised. Likewise if the anode potential (μ_A) is above the LUMO (lowest unoccupied molecular orbital) level, then it will reduce the electrolyte (Figure 5.19). The electrolyte is stable only if the electrochemical potential of both the anode and cathode remains within the stability window of the electrolyte (E_g) which is defined as the difference between the energy of the HOMO and LUMO. The TiS_2 and $LiCoO_2$ cathodes both lie within the electrolyte's stable range but the electrode potential of lithium and graphite anodes is higher than the LUMO. This incompatibility was overcome by the kinetic stability conferred by the solid electrolyte interface (SEI) which acts as a barrier protecting the anode from direct contact with the electrolyte [see Section 5.4.2.5].

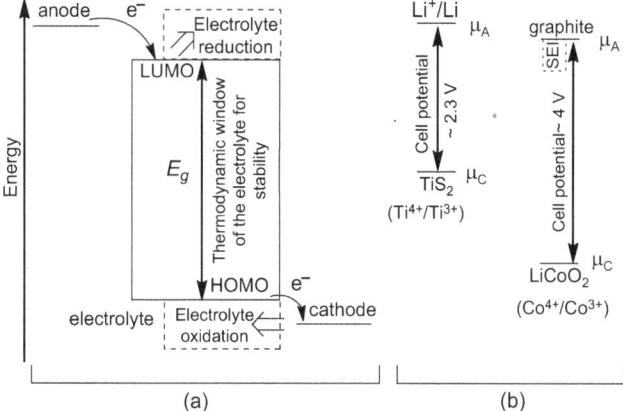

(a)   (b)

FIGURE 5.19   Integration of electrodes and electrolyte. (a) Schematic energy diagram of potential side reactions between electrodes and electrolyte. (b) Matching electrochemical potentials of the electrodes in Li-TiS_2 and graphite-$LiCoO_2$ cells to the working window E_g of the electrolyte. Electrochemical potentials μ_A and μ_C in solid state chemistry and physics are commonly quoted in the energy units of electron volts (eV) and 1 eV is equivalent to 96.48 kJ/mol.

5.4.2.4 A Practical Innovation

A lithium-ion battery starts its life in a state of full discharge with all its lithium ions intercalated within the $LiCoO_2$ cathode. Assembly of a cell in a discharged state is practical for a rechargeable battery and was another novel aspect of the development of lithium-ion technology. During the charging process lithium ions are extruded from the $LiCoO_2$ structure by migrating from one octahedral site to another via an intermediate tetrahedral site. As lithium ions are removed from the lattice the electrostatic repulsions between adjacent negatively charged CoO_2 layers increase causing expansion along the c crystallographic axis. Complete removal of lithium ions leads to a shift from the triple-layer stacking to a single-layered arrangement with the oxygen layers rearranging to give a hexagonal close-packed structure and such phase transitions are detrimental to electrochemical performance. Around half of the lithium ions could be reversibly cycled from $LiCoO_2$ without causing changes to its structure. A $LiCoO_2$ cathode and lithium metal anode battery can give an output voltage near 4 V which was almost twice that obtained with a metal sulphide cathode.

5.4.2.5 The Anode Revisited

A delithiated intercalation anode was required to combine with the lithiated cathode as the cells are assembled in the discharged state. Graphite was the front runner from all the anode candidates considered as it was known to trap a variety of guest species including alkali metal ions in the interlayer sites and is an electrical conductor. Graphite has a layered structure of carbon atoms with strong covalent bonding within each layer but weak van der Waals forces between adjacent layers (Figure 5.20a). Each carbon atom has four valence electrons and forms three σ bonds to the adjacent carbon atoms, generating an aromatic ring, leaving one electron to participate

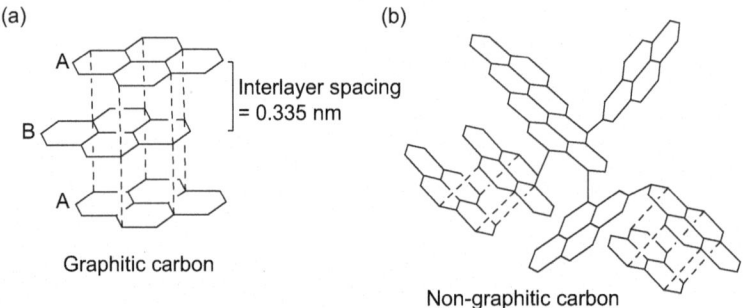

FIGURE 5.20 Graphitic and non-graphitic forms of carbon. (a) Part of the layered lattice structure of graphite. The carbon atoms form layers of hexagonal rings which are stacked in an ABAB (shown) or ABCABC sequence with a ratio of 70:30 in the natural mineral. (b) Petroleum coke, a by-product from oil refineries, can be converted into a graphite-type material consisting of randomly oriented graphite sheets joined by amorphous carbon regions. Both the natural and synthetic forms of graphite are used as anodes in lithium-ion batteries. Historically, petroleum coke served as the anode in the first commercial lithium-ion batteries as is worked well with the propylene carbonate electrolyte.

in delocalised π bonding. The molecular π orbitals extend over each layer and the energy gap between the fully occupied molecular orbital and the vacant antibonding molecular orbital is very small which enables the electrical conductivity parallel to the layers to approach that of a metal, i.e. in this respect graphite can be considered a semi-metal.

The chemical route to intercalating alkali metal ions into graphite involves directly reacting graphite with the molten metal or its vapour. In this reaction the alkali atom transfers its electron to the delocalised π orbitals of the graphene sheet and the cation enters the interlayer region. Fully lithiated graphite has one lithium per six carbon atoms giving the formula LiC_6. Problems emerged in intercalating Li^+ ions into graphite electrochemically as the electrolyte reacted with the lithiated graphite. Lithiated carbon is a powerful reducing agent and intercalated propylene carbonate electrolyte molecules decomposed leading to delamination and collapse of the graphite structure. Replacing propylene carbonate, a traditional component of non-aqueous electrolytes, with ethylene carbonate, which lacks the methyl substituent on the five-membered ring (see Figure 5.16), was a major step forward in preventing the reductive decomposition reaction. Ethylene carbonate has a higher melting temperature compared to the propylene compound and forms a stable SEI layer on the graphite. This layer prevents the direct contact of the graphite particles with the electrolyte and kinetically suppresses electrolyte decomposition and inhibits solvent co-intercalation whilst allowing migration of lithium ions further into the graphite particles. The SEI forms during the first cycles of a battery and contains compounds resulting from reactions between electrolyte molecules and lithium ions which leads to some depletion of lithium ions from the system. As lithium-ion batteries are assembled in the discharged state this process is managed and manipulated in the factory through carefully formulated battery cycles and electrolyte additives. Safety features are also built into lithium-ion batteries to prevent the complete discharge and over-charging and to prevent overheating.

5.4.2.6   Later Developments and the Current Focus of Research

The original lithium-ion battery has been changed and improved over the years and research continues apace as seen by the ever-increasing number of scientific papers published in the field of intercalation batteries (Figure 5.21). The new electrolyte formulation consisting of $LiPF_6$ in an ethylene carbonate/dimethyl carbonate mixture was introduced in 1993 and increased the energy density of the battery whilst in the early 2000s polymer electrolyte-based lithium-ion batteries became commercially available. Replacing the flammable electrolyte by suspending the lithium salt in a conducting polymer gel such as polyethylene oxide or polyacrylonitrile rather than a nonaqueous solvent enhances safety and the battery can also be made lighter, thinner and more flexible.

In lithium-ion batteries $LiCoO_2$ is still one of the best cathode choices for portable electronic devices where battery volume is at a high premium but for larger applications, such as vehicle electrification or grid storage, other factors come into play. Nickel-manganese-cobalt (NMC) cathodes, $Li[Ni_x Mn_y Co_z]O_2$ $(x + y + z = 1)$, are widely used in electric vehicles partly to reduce the reliance on cobalt, a scarce and toxic element, in the battery and partly because of safety concerns surrounding

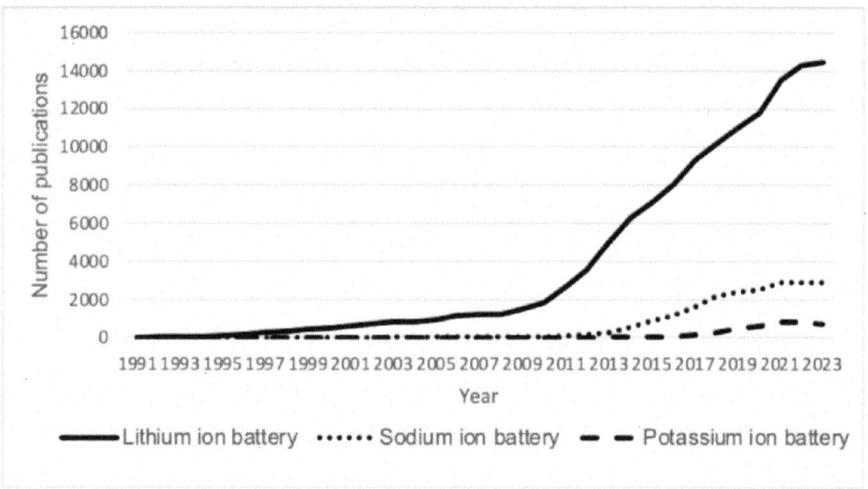

FIGURE 5.21 Research publications on alkali ion batteries since 1991 when Sony launched the first petroleum coke/LiCoO$_2$ battery. (Data from Web of Science, a database of articles published in the highest impact journals worldwide, including open access journals. Date accessed: 5 September 2024.)

LiCoO$_2$. At high temperatures (~200°C) LiCoO$_2$ can decompose and produce oxygen (O$_2$) which, when combined with a flammable electrolyte, can cause fires. Lithium iron phosphate, LiFePO$_4$, batteries are being revisited as another option for electric vehicles as they eliminate the need for cobalt and are thus more environmentally friendly than those with NMC cathodes but have a lower energy density and so take up more space. The uneven distribution of the world's lithium resources is also driving the development of alternative and more sustainable intercalation battery chemistries based on other ions such as Na$^+$ or K$^+$.

Development of LCDs and lithium-ion batteries has its origins in the oil crisis of the 1970s and the consequent need to reduce reliance on fossil fuel-based energy. Maturity of the fields in the intervening years, particularly in regard to battery storage, is facilitating the move towards decarbonising energy generation and helping to combat climate change.

5.5 PHOTOGRAPHIC CHEMISTRY

Digital cameras and camera lenses integrated into digital devices are commonplace in the 21st century and are the means by which the vast majority of photographic images are produced today. Photography requires light and in digital photography light reflected or emitted from objects is captured and stored electronically. The image is available to view immediately and can be transferred to a computer for digital manipulation. Before the advent of digital methods photography used film containing light-sensitive chemicals to capture a latent (invisible) image. The latent image is made visible in a development step with the resulting image 'fixed' to prevent any further action by light (see Section 5.5.1). Most 'analogue' photographs until the mid-20th century were black-and-white (monochrome) in which the colours of the original scene were translated into shades of grey. Colour photography only became routine and affordable when dye-based emulsion systems were introduced by a number of companies. Developing the latent image was a complicated process with further dyes being involved in some of the (many) processing steps before the colour photograph was produced. Traditional colour film photography has now largely been superseded by digital technology whereas black-and-white film remains a favoured medium amongst some photographers.

The action of light on silver salts was recognised in the early 18th century and around 1840 several photographic processes became commercially available. Two of these, the Daguerreotype (invented by Louis Daguerre) and the calotype (invented by William Henry Fox Talbot), were based on the light sensitivity of silver halides to capture a latent image. The calotype gave a paper negative which was used to print a positive image whilst the Daguerreotype gave a positive image directly. Both processes became obsolete when the wet collodion process was developed by Frederick Scott Archer in 1851. This method used dispersions of silver halides in nitrocellulose and either negative or positive images could be produced. The wet collodion process was cheaper than its predecessors, did not require the toxic mercury developing reagents associated with the Daguerreotype and exposure times were shorter. Although there were drawbacks with the wet collodion process it dominated the photographic market for the next 30 years. A method to turn nitrocellulose into a flexible film base in 1888 led to the development of roll film and paved the way for cinematography. Advances in optics and photographic chemistry reduced exposure times further; cameras became lighter and more portable and in the 20th century 35-mm plastic-backed roll film giving 24 and 36 images was widely adopted by both amateurs and professionals.

The introduction of digital photography made the practice of creating and reproducing images quicker and easier and more accessible to more people. Analogue photography in comparison is a slower process and, given the limited number of exposures on a film, greater consideration is generally given to the picture taken. Processing the film is normally done commercially and there is often the option of obtaining the images digitally as well as in the print form. Printing black-and-white negatives oneself is however a satisfying and creative experience and simple techniques such as dodging or burning-in areas can improve the original image. Whilst this procedure does require access to a darkroom the two photographic experiments in Chapter 6 do not. Both these experiments give scope for individual creative endeavours in addition to being practical examples of redox chemistry.

5.5.1 BLACK-AND-WHITE PHOTOGRAPHY

The key compounds in analogue film photography are light-sensitive silver halides, particularly silver bromide, AgBr. In black-and-white photography, microscopic grains of AgBr, each about 0.1–1 µm in diameter, are embedded in a gelatine emulsion which is supported on a transparent plastic base (Figure 5.22). The crystals of AgBr are not perfect; they contain flaws, cracks and impurities and it is these defects in the crystal lattices which enable the latent image to form. The invisible latent

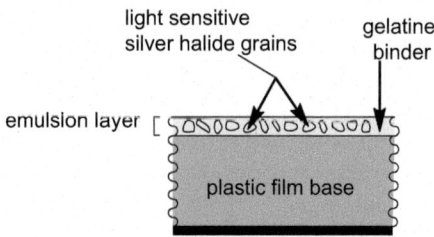

FIGURE 5.22   Composition of photographic black-and-white film.

image created when the picture is taken with the camera is then converted into the visible and permanent (negative) image by the processing solutions in the laboratory (see below). The negative image has dark and light areas which are the reverse of the picture taken with the camera, i.e. areas in the original scene that were relatively dark appear light and areas which were bright appear dark. To obtain the image that reflects the original photographed scene the tonal values of the negative need to be reversed and this is achieved in the printing process which is a separate, although similar, development procedure to that used to create the negative.

5.5.1.1   Exposure and Formation of the Latent Image

The sensitive medium on the photographic film is the microscopic crystals of AgBr suspended in a gelatine matrix. When the camera shutter opens and light strikes the film emulsion a photon reacts with a silver halide crystal causing an electron to be ejected from the valence shell of the bromide ion to form a bromine atom (Figure 5.23). In this photoelectric process which ejects an electron from the bromide ion the energy of the electromagnetic radiation in converted into kinetic energy of the electron. This high-energy electron then moves around the crystal until it is trapped by an imperfection. Imperfections, known as sensitivity spots, arise during the preparation of the photographic emulsion and are thought to involve the formation of silver sulphide. After several photons have reacted with AgBr crystals, an accumulation of the ejected electrons is created around the local imperfection. This local accumulation of electrons generates a negative electric field which attracts positive interstitial silver ions located on Frenkel defects and are able to migrate through the lattice structure. Silver ions then combine with the free electrons to give metallic silver atoms. The tiny amount of metallic silver (about 1 atom for every 10 million silver bromide formula units) that forms within the crystal grain after exposure is known as the latent image. This image is invisible to the naked eye and the film appears unchanged, remaining dull and opaque. The latent image is converted into a visible image by the process of development.

5.5.1.2   Development and Production of a Visible Image

Development works as an amplification process of the latent image and involves treating the exposed film with a reducing agent such as hydroquinone. Hydroquinone causes specks of silver to form in areas that were exposed to light and the quantity of silver formed is directly proportional to the amount of exposure. Grains of silver bromide are reduced to metallic silver but, as this reaction is catalysed by silver, only

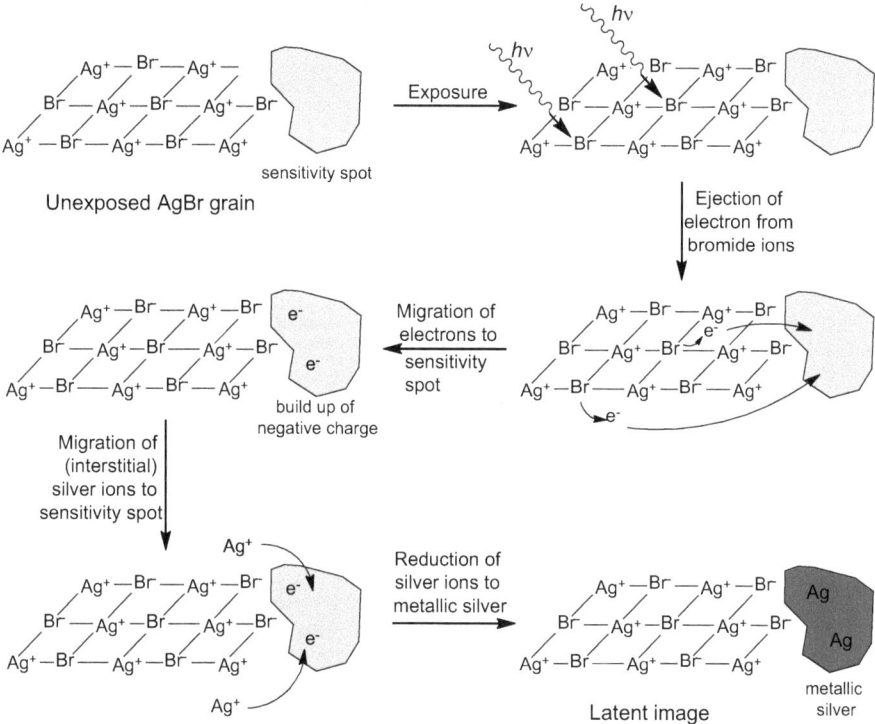

FIGURE 5.23 Formation of the latent image. Silver bromide has a crystal structure similar to NaCl (see Figure 5.11a) in which each Ag^+ ion is surrounded by six Br^- ions and vice versa. Note that the ionic bonds are often represented as straight lines in three-dimensional structures or in fragments thereof such as that shown above, to emphasise the relationship between the oppositely charged ions. The radius of an Ag^+ ion is 115 pm which is much smaller than that of Br^- at 196 pm and this can cause Frenkel defects in the crystal lattice. A Frenkel defect is a point defect in which the smaller Ag^+ ion migrates from an octahedral site to a (normally unoccupied) tetrahedral hole in the close-packed structure and in this position the Ag^+ is referred to as an interstitial ion.

those grains of silver bromide that contain microscopic deposits of metallic silver in the latent image are reduced (Figure 5.24). The film is left in contact with the developing solution for a time sufficient to build up the image to the required macroscopic density, i.e. to reveal the visible image. The length of time the film is left in the developing solution is critical as the hydroquinone, if left in contact with the emulsion, will eventually reduce all the silver bromide crystals. Development is therefore a kinetically controlled process which magnifies the difference between reduction of sodium bromide crystals which have been exposed to light and those that have not.

5.5.1.3   Stopping and Fixing; Making the Visible Image Permanent

When the film is taken out of the developing tank the reduction reaction continues, so a stop-bath is used to halt the process. Developing requires an alkaline environment to work so a stop-bath is a weak acid, usually ethanoic (acetic) or citric acid.

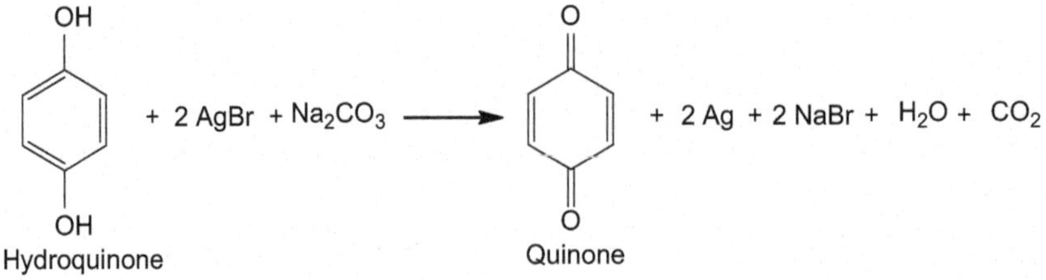

FIGURE 5.24 Reduction of silver bromide with hydroquinone in the developer solution. The Ag$^+$ ions in AgBr are reduced to Ag metal and the hydroquinone is oxidised to quinone. The reduction reaction requires alkaline conditions with sodium carbonate (Na_2CO_3) being the most widely used alkali. Development time is also influenced by temperature which is usually 18°C–20°C.

The areas of the film containing silver halide crystals which were not exposed to light are still light sensitive and these need to be flushed out of the emulsion. This is the function of the fixing agent – it dissolves the silver halides into a solution and now being soluble, they can be flushed out with the fixer agent. A traditional fixing agent is sodium thiosulphate (sometimes known as 'hypo'), $Na_2S_2O_3 \cdot 5H_2O$. The thiosulphate ions are adsorbed onto the surface of the grains and form $Ag_2S_2O_3$, a sparingly soluble salt, which is converted into the water-soluble salt, sodium silver thiosulphate, $Na_3[Ag(S_2O_3)_2]$, in excess hypo (Figure 5.25). The emulsion and the silver ions which have already been reduced to metallic Ag are not affected by the fixer. Finally, the film is washed thoroughly to remove all traces of fixer and then dried.

The film is now insensitive to light and can be looked at in daylight. As the film base is transparent, emulsion parts which did not receive any light appear transparent because unexposed silver bromide has been washed out. In contrast, parts of the film that had received light have been converted to metallic silver and appear opaque. As the opaque parts match the light parts of the original scene and the transparent parts match the dark parts, i.e. the light and shade regions of the picture are reversed, the image is referred to as the negative of the original picture taken.

5.5.1.4 Printing: Making a Positive from a Negative

Negatives from a 35-mm roll of film give images that are 24 mm × 36 mm (height × width) and are generally enlarged when printing to enable the detail in the original photographed scene to be seen.[9] The negative image is projected using an enlarger onto light-sensitive photographic paper which is paper coated with an emulsion of silver bromide in gelatine. Shining light through the negative and projecting the image onto the light-sensitive paper means that transparent and opaque parts of the negative will let through more and less light respectively and thus reproduce the dark and light shades of the original photographed scene. Steps in the printing process therefore mirror those above, i.e. exposure, developing, stopping and fixing but, as black-and-white photographic paper is less sensitive to light than the original 35-mm film, printing can be carried out in a darkroom under a (red or amber) safelight. Exposure times are therefore also longer (but usually less than a minute) than when taking the original picture which gives an opportunity to give some areas of the print more or less light (using techniques of 'burning in' and 'dodging') and afford the finished picture with greater detail.

$$Na_2S_2O_3 \text{ (aq)} + 2\,AgBr \text{ (s)} \longrightarrow Ag_2S_2O_3 \text{ (s)} + 2\,Na^+ \text{(aq)} + 2\,Br^- \text{(aq)}$$

and in excess hypo: $Ag_2S_2O_3 \text{ (s)} + 3\,Na_2S_2O_3 \text{ (aq)} \longrightarrow 2\,Na_3[Ag(S_2O_3)_2] \text{ (aq)}$

FIGURE 5.25   Fixing reactions with hypo (sodium thiosulphate).

Other creative techniques can be used to further manipulate the black-and-white printed image and one of these, copper toning, is explored in Experiment 6.4.2. Black-and-white negatives can however be printed using cyanotype paper and this technique is explored in the second of the two photographic experiments in Chapter 6.

NOTES

1 Symbols that represent a pictorial concept.
2 Mixture of brassicalactone, phytol and 6,10,14-trimethylpentadecan-2-one in a 16:31:10 ratio.
3 Indicated by the female sitting still, folding her wings and accepting a male's advances as opposed to a refusal posture with wings spread horizontally and a vertically lifted abdomen or flying away when rejecting courtship. The females control mating; male butterflies are incapable of forcing mating on females and so female receptivity is a necessary prerequisite for mating.
4 These findings were obtained from a series of elegant experiments in which male *P. napi* butterflies were fed with labelled L-Phe and the labelled methyl salicylate metabolite was identified in their abdomens. In these experiments some of the ^{12}C atoms in L-Phe had been replaced by the ^{13}C isotope thus enabling the presence of the labelled methyl salicylate to be identified through mass spectrometry. Other experiments involved feeding several of the aromatic intermediates which are en route to methyl salicylate from L-Phe and tracking them through to the anti-aphrodisiac using deuterium labelling, i.e. some of the compounds fed to the butterflies incorporated deuterium atoms (^2H) instead of hydrogen (^1H) atoms. Neither the larvae nor the adult females were shown to produce the anti-aphrodisiac.
5 In practice the chain extending unit is a malonate group, $HO_2C–CH_2–CO–SCoA$, a more reactive form of acetate which is formed through a reaction between acetyl-SCoA and CO_2. Acetyl-SCoA and malonyl-SCoA themselves are not involved in the condensation reaction but are converted into enzyme-bound thioesters. Condensation occurs with simultaneous loss of the extra carboxyl group in the malonate unit to give the butanoyl thioester $CH_3–CO–CH_2–CO–SEnz$ and the biosynthetic process then continues with reduction of the carbonyl group. Absence of the reduction step leads to the polyketide class of natural product.
6 Electrochemistry determines the voltage (electrical potential difference) which is the driving force to push electrons around a circuit. Alkaline AA and AAA batteries with the same chemistry have the same voltage but the larger AA battery contains more of the chemicals for the reactions, i.e. they have a larger energy store and so can deliver more current over a longer period of time. The 12 V lead-acid batteries in cars are made from six 2 V cells linked together. Historically, the term battery referred to a series of voltaic cells joined together; however, the usage has evolved to include devices composed of a single cell.
7 By convention, electrode potentials are always quoted in terms of the reduction equation, $M^{n+} + n\,e^- \rightarrow M$, with (larger) positive values indicating that this is the direction of the spontaneous reaction. The spontaneous reaction in the reverse direction, i.e. oxidation: $M \rightarrow M^{n+} + n\,e^-$, is indicated by standard electrode reduction potentials with (large) negative values. The Li^+/Li has a reduction potential of –3.05 V with the overall cell potential being a measure of

the difference between the electrode potential of anode and cathode (see also Figure 5.14b). The combination of elements at the anode and cathode which would theoretically give the largest cell potential of 5.94 V (in an aqueous environment) is a reduction of fluorine $\left[E^{\circ}_{red}\left(\frac{1}{2}F_2/F^- \right) = 2.89 \text{ V} \right]$ at the cathode and oxidation of lithium $[E^{\circ}_{red}\left(Li^+/Li = -3.05 \text{ V} \right]$ at the anode, but the extreme reactivity of elemental fluorine makes this impractical.

8   In network solids, which have large numbers of atoms the discrete energy levels of individual atoms spread to become a band of allowed energy made up of closely spaced energy levels each capable of containing two electrons. Electrons responsible for chemical bonding form the valence and lower energy band of the solid and an upper energy conductance band has the capacity to accommodate electrons. Electrons in the conduction band are relatively free and can take part in electrical conduction.

9   35 mm is the total height of the film and includes the perforations along each side which engage with sprockets in the camera and facilitates the loading and advancing of the film through the camera.

10 When the liquid crystalline phase occurs as a result of temperature change it is known as a thermotropic liquid crystal and these are the type of molecules used in LCDs. Another type of liquid crystal exists, the lyotropic liquid crystal, which is based on the self-assembly of amphiphilic molecules in a solvent, typically water, and the phase transition is a function of both temperature and concentration. This type of liquid crystal is abundant in nature; lipid bilayers of cell membranes for example exhibit a mesophase as do some proteins, such as spider silk, when stored in the arachnid's spinning gland.

FURTHER READING

Academic papers detailing the Lepidoptera pheromone research and review articles on lithium-ion batteries and liquid crystal displays are listed below with more general background material relating to these topics being found in the specialist books. No specific literature is given for photography as this is a mature subject and there is a large amount of material to consult on all aspects of the field both in print and online.

BOOKS

Asher, J., Warren, M., Fox, R., Harding, P., Jeffcoate, G. and Jeffcoate, S. (2001) *The Millennium Atlas of Butterflies in Britain and Ireland*. Oxford: Oxford University Press.

Burrows, A., Holman, J., Lancaster, S., Overton, T., Parsons, A., Pilling, G. and Price, G (2021) *Chemistry³: Introducing Inorganic, Organic and Physical Chemistry*. 4th Ed. Oxford: Oxford University Press.

Mann, J. (1994) *Chemical Aspects of Biosynthesis*. Oxford: Oxford University Press.

ARTICLES

PHEROMONES

Andersson, J., Borg-Karlson, A.-K., Vongvanich, N. and Wiklund, C. (2007) Male Sex Pheromone Release and Female Mate Choice in a Butterfly. *The Journal of Experimental Biology*. [Online] 210 (6). pp. 964–970. Available from: https://doi.org/10.1242/jeb.02726. [Accessed: 11th June 2024].

Bradshaw, J. W. S., Baker, R. and Lisk, J. C. (1983) Separate Orientation and Releaser Components in a Sex Pheromone. *Nature*. [Online] 304. pp. 265–267. Available from: https://doi.org/10.1038/304265a0. [Accessed: 11th June 2024].

Fatouros, N. E., Huigens, M. E., van Loon, J. J. A., Dicke, M. and Hilker, M. (2005) Butterfly Anti-Aphrodisiac Lures Parasitic Wasps. *Nature*. [Online] 433. p. 704. Available from: https://doi.org/10.1038/433704a. [Accessed: 15th June 2024].

Hicks, B. J., Leather, S. R. and Watt A. D. (2008) Changing Dynamics of the Pine Beauty Moth (*Panolis flammea*) in Britain: The Loss of Enemy Free Space? *Agricultural and Forest Entomology*. [Online] 10. pp. 263–271. Available from: https://doi.org/10.1111/j.1461-9563.2008.00382.x. [Accessed: 11th June 2024].

Huigens, M. E., Pashalidou, F. G., Qian, M.-H., Bukovinszky, T., Smid, H. M., van Loon, J. J. A., Dicke, M. and Fatouros, N. E. (2009) Hitch-Hiking Parasitic Wasp Learns to Exploit Butterfly Antiaphrodisiac. *Proceedings of the National Academy of Sciences of the United States of America*. [Online] 106 (3). pp. 820–825. Available from: https://doi.org/10.1073/pnas.0812277106. [Accessed: 15th June 2024].

Mozuraitis, R., Murtazina, R., Zurita, J., Pei, Y., Ilag, L., Wiklund, C. and Borg Karlson, A. K. (2019) Anti-Aphrodisiac Pheromone, a Renewable Signal in Adult Butterflies, *Scientific Reports*. [Online] 9. Article no. 14262. Available from: https://doi.org/10.1038/s41598-019-50838-1. [Accessed: 5th June 2024].

Yildizhan, S., van Loon, J., Sramkova, A., Ayasse, M., Arsene, C., ten Broeke, C. and Schulz, S. (2009) Aphrodisiac Pheromones from the Wings of the Small Cabbage White and Large Cabbage White Butterflies, *Pieris rapae* and *Pieris brassicae*. *ChemBioChem*. [Online] 10 (10). pp. 1666–1677. Available from: https://doi.org/10.1002/cbic.200900183. [Accessed: 15th June 2024].

LIQUID CRYSTALS

Dierking, I. (2014) Chiral Liquid Crystal: Structures, Phases, Effects. *Symmetry*. [Online] 6. pp. 444–472. Available from: https://doi.org/10.3390/sym6020444. [Accessed: 13th November 2024].

Hilsum, C. (2010) Flat-Panel Electronic Displays: A Triumph of Physics, Chemistry and Engineering. *Philosophical Transactions of the Royal Society A*. [Online] 368. pp. 1027–1082. Available from: https://doi.org/10.1098/rsta.2009.0247. [Accessed: 2nd July 2024].

BATTERIES

Goodenough, J. B. (2019) The Pathway to Discovering Practical Cathode Materials for the Rechargeable Li+-ion Battery. *Nobel Lecture*. [Online] Available from: https://www.nobelprize.org/prizes/chemistry/2019/goodenough/lecture/. [Accessed: 10th June 2024].

Xie, J. and Lu, Y.-C. (2020) A Retrospective on Lithium-Ion Batteries. *Nature Communications*. [Online] 11. p. 2499. Available from: https://doi.org/10.1038/s41467-020-16259-9. [Accessed: 13th August 2024].

PHOTOGRAPHY

Royal Photographic Society (2024) [Online] Available from: https://www.rps.org. [Accessed: 9th December 2024].

6 Experiments

6.1 INTRODUCTION

The five sets of experiments in this chapter involve several compounds which feature in Chapters 4 and 5 on the chemistry of colour and communication. Some experiments focus on compounds with a very rich chemical history whilst others focus on molecules which have come to prominence in more recent years. One series of experiments is centred around using natural dyes to colour fabrics and is designed to be partly investigative. Reduction of chloroauric acid, $HAuCl_4$, to form gold nanoparticles is the focus of a second set of experiments but there is also scope for further exploration of this reduction reaction using the procedures given in this chapter as a starting point. Two types of photographic-based processes involving the redox chemistry of iron compounds – cyanotype, also known as the blueprint process, and copper toning of black-and-white prints – form a third group of experiments. The 'blue' in the blueprint process and the red-brown colouration in the copper toning process are due to the formation of Prussian blue, $Fe_4[Fe(CN)_6]_3$, and copper(II) hexacyanoferrate(II), $Cu_2[Fe(CN)_6]$ respectively, both of which are complexes of the octahedrally coordinated iron(II) ion, $[Fe(CN)_6]^{4-}$. Tris(8-hydroxyquinoline) aluminium(III), Alq_3, and cholesteryl benzoate were key compounds in the development of organic light-emitting diodes (OLEDs) and liquid crystalline materials respectively and their syntheses from readily available starting materials are the focus of two preparative experiments. Videos of some of the procedures, marked with an asterisk (*) in the text, are available in the online material.

6.1.1 HEALTH AND SAFETY

Step-by-step procedures are given for each of the experiments together with a list of the chemicals, equipment and apparatus required. All of the experiments involve weighing out chemicals and many of the procedures involve heating reaction mixtures. A stirrer hotplate is used for heating reaction mixtures in the procedures given in this chapter but other electrically powered heat sources such as a water bath are suitable alternatives. Whilst Bunsen burners can be used to heat aqueous solutions they must not be used in experiments where there is the possibility of flammable vapours coming into contact with the flame and causing a fire. Hazardous chemicals and operations are identified at appropriate points of the experimental procedures using the word **CARE in capitals and bold type** followed by *specific warnings and recommended precautions in bold and italics*. Note that a hazard is the potential of a substance or activity such as heating to cause harm whilst a risk is the likelihood of harm being caused by a hazard. The potential hazards in the experiments within this chapter can be minimised by conducting the procedures on a small scale to lessen the adverse effects of any accident. Furthermore, small-scale reactions also reduce the amount of waste materials generated and decrease the cost of the experiments.

DOI: 10.1201/9781003562313-6

The photographic experiments and those involving dyestuffs utilise chemicals which are at the lower end of the hazardous scale and are carried out in aqueous solutions. As a result it may be possible, with the proper regard to safety considerations, to carry out these experiments in a less formal laboratory setting such as that found in a domestic garage. However, the Alq_3 and cholesteryl benzoate syntheses which use flammable solvents and more hazardous reagents must be carried out in a fume cupboard in a proper chemical laboratory under the guidance of a qualified chemist.

Whilst every effort has been made to ensure the experimental procedures in this chapter are accurate and that potential hazards have been identified, it is ultimately the responsibility of the individuals carrying out the experiments to do so in a safe manner. These individuals must complete a thorough risk assessment of the chemicals and procedures before any practical work is started and record this information in their laboratory notebook. Safety Data Sheets for the chemicals are available online and can be obtained by typing the chemical name and a chemical supplier such as Sigma-Aldrich or Merck into an appropriate search engine. Personal protective equipment (PPE) consisting of safety glasses and a laboratory coat must be worn at all times when carrying out the experiments and gloves may be required when handling some chemicals. A list of essential rules for laboratory safety is given below.

Always
- Wear eye protection and a fastened laboratory coat when undertaking chemistry experiments. Note that contact lenses should not be worn when carrying out the experiments as chemicals can get under the lens and damage the eye before the lens can be removed.
- Dress sensibly; long hair must be tied back and proper shoes must be worn. Note that open-toed sandals do not provide adequate protection for feet.
- Read the instructions carefully before starting any experiment.
- Wash hands thoroughly with soap when experimental work has been completed and before leaving the laboratory.
- Handle all chemicals with great care.
- Carry out practical work standing up.
- Keep work spaces tidy to help prevent accidents and spillages.

Never
- Eat or drink in the laboratory.
- Inhale, taste or sniff chemicals.
- Run, fool around or distract others in the laboratory.
- Work alone.

Note that the authors cannot be held responsible in any way whatsoever for mishaps incurred during experimentation.

6.1.2 CHEMICALS, EQUIPMENT AND APPARATUS

Dyeing fabrics with natural products and photographic processing are both popular hobbies and the chemicals for these procedures can be sourced from online suppliers. Chloroauric acid ($HAuCl_4$) is a reagent in the nanoparticle experiments and, as this compound is also used in gold toning processes in photography, it may be available through photographic suppliers. Photographic suppliers also stock some equipment and apparatus for chemical processing procedures such as cyanotype and toning experiments. The small-scale nature of the experiments involves using a balance that can accurately weigh out solid chemicals to the nearest 0.01 g and access to a two decimal place balance is essential. This type of equipment and glassware such as beakers, thermometers and measuring cylinders can be obtained through various chemical suppliers. The purchase of chemicals for the syntheses of tris(8-hydroxyquinoline) aluminium(III) and cholesteryl benzoate may be restricted to individuals who are employed in a recognised research, educational or other approved organisation.

The step-by-step written instructions for the experiments aim to be comprehensive but setting up equipment and apparatus to carry out chemistry reactions can be daunting to individuals with little prior experience of laboratory work. Video recordings of some experiments are provided in the online material to show those who are less confident in their laboratory skills how to organise and carry out the different steps of the written instructions. Furthermore, these resources may be useful for individuals to visualise the experiments if they are not able to perform the reactions in person in a laboratory. Other resources to guide individuals in the practice of laboratory chemistry are given at the end of this chapter in the *Further Reading* section.

It is good practice for anyone carrying out experiments, even if following a written procedure, to keep a record of their work in a notebook. Record the masses and volumes of all the starting materials, the temperatures and reaction times etc., note down observations during the experiment and clearly indicate any deviations from the written procedure. In the investigative experiments this information will help to decide the direction of subsequent procedures. Recording numerical results such as the mass of product and analytical data, e.g. melting point or R_f values, from preparative experiments helps assess the success of the reactions.

6.2   DYEING FABRICS WITH NATURAL PRODUCT DYES

Natural product dyes have been used for centuries to impart colour to fabrics. Reds were often derived from madder plants (*Rubia tinctorum*) or cochineal insects (*Dactylopius coccus*) whilst yellow and blue colours often came from weld (*Reseda luteola*) and indigo (*Indigofera tinctoria*) plants respectively. Overdyeing weld-coloured materials with indigo was often a route to generate a green colour. The chemical identities of the main molecules imparting colours are now known to be alizarin (red), carminic acid (red), luteolin (yellow) and indigo (blue) (Figure 6.1). In this series of experiments the powdered dried extracts of the plant or animal source are used as the dyestuffs and other organic compounds will be present in the extracts and may contribute to the colours obtained. For readers also interested in botany the dye plants can be grown in the garden and then either specific parts (flowers, leaves, stems, roots) or the whole plant harvested and processed to generate the dyestuff. Detailed information on this approach is given in the *Further Reading* section at the end of this chapter.

The experimental procedures focus on the dyeing of natural fibres (wool, silk and cotton) with the classic natural dyes cited above. Wool and silk are protein (animal) fibres whereas cotton is a cellulose (plant) fibre and these two types of materials are

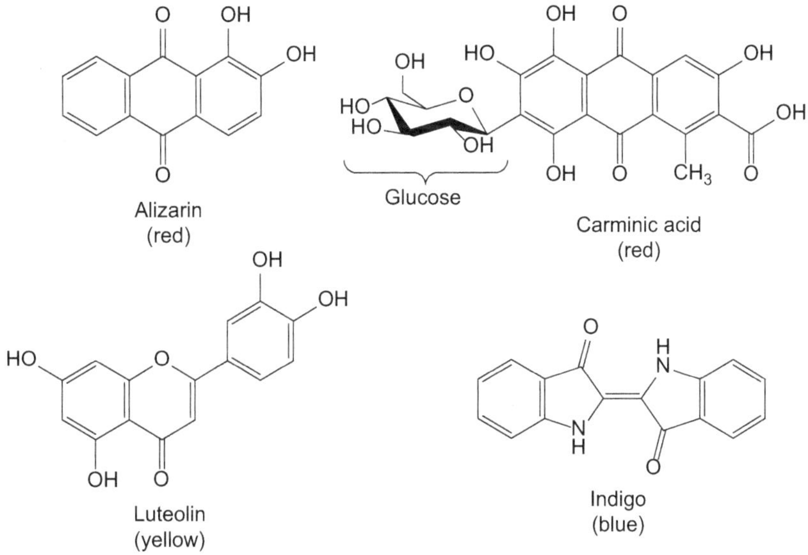

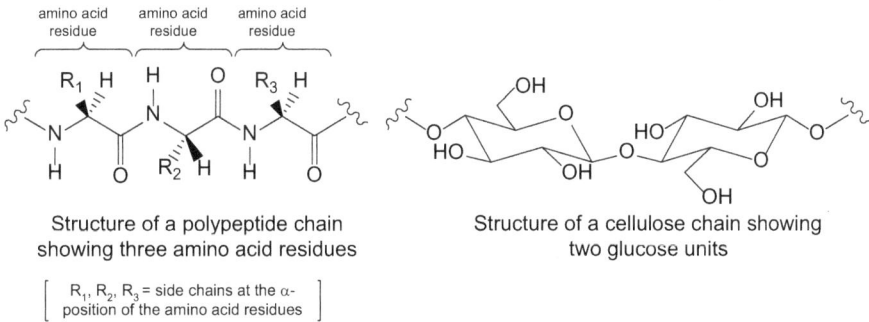

FIGURE 6.1 Chemical structures of organic dye molecules from madder, cochineal, weld and indigo.

Structure of a polypeptide chain showing three amino acid residues

Structure of a cellulose chain showing two glucose units

R_1, R_2, R_3 = side chains at the α-position of the amino acid residues

FIGURE 6.2 Structure of polypeptide/protein and cellulose chains. Chains can be built from many hundreds of amino acids or glucose molecules.

biological polymers formed of chains of amino acids and glucose units respectively (Figure 6.2). Glucose contains many hydroxyl (OH) groups in the molecule whereas the amino acids which make up the polypeptide chains have a range of functional groups including acidic, basic, aromatic, hydroxyl and thiol (SH) moieties and so animal and plant fibres can react differently with dye molecules. Once the basic dyeing procedures have been mastered, and the end results of dyeing wool, silk and cotton evaluated, there is tremendous scope for investigating other plant dyes and fibres.

Mordanting fabrics facilitates the formation of chemical bonds between dyestuffs and fibres which results in more vibrant colours and procedures for this process are also given. Mordants are either metal salts or tannins (plant-based polyphenol molecules) and the effect of different mordants on the take-up of dye into fibres is also an area which can be explored further.

Indigo, unlike madder, weld and cochineal extracts, does not require a mordant to dye either protein or cellulose materials as it dyes through a physical attachment process rather than forming chemical bonds to the fibres. Indigo is first reduced to the soluble and colourless or pale yellow substance (leucoindigo) which penetrates into the fibres during the immersion step. When the fibres are removed from the solution and exposed to air, the leucoindigo molecules are oxidised back into the insoluble indigo form and are thus physically trapped, which causes the treated materials to be coloured blue (see Chapter 4, Figure 4.15 for the reaction scheme which illustrates the reduction and oxidation steps).

The wool, silk and cotton fibres and fabrics used in the following procedures were obtained from craft suppliers or haberdashery stores and were either a naturally pale colour or had been bleached white. Silk and cotton pieces in the procedures below were approximately 6 cm × 2.5 cm and the wool (double knitting thickness) and cotton yarns were 25 cm in length. The fibres and fabrics are collectively referred to as 'fibres' in the procedures below.

6.2.1 MORDANTING FIBRES WITH METAL SALTS

Materials

Either	1.00 g	not a hazardous substance
alum (potassium aluminium sulphate), $KAl(SO_4)_2 \cdot 12H_2O$		
or		
aluminium acetate $Al(OCOCH_3)_3$	1.00 g	**skin and eye irritant**

deionised water (*if this is not available use rain or tap water*)

six pieces each of the wool, silk and cotton fibres and fabrics

Equipment: hotplate, thermometer (0°C–100°C), 100 cm³ measuring cylinder, 250 cm³ beaker (×2), spatula, tweezers

---

Procedure

1. Weigh out 1.00 g of the mordant (alum or aluminium acetate) into a 250 cm³ beaker and add 100 cm³ of water. Stir the mixture until the solid has completely dissolved.
2. Weigh the fibres in a 250 cm³ beaker. Add approximately 100 cm³ water and leave the fibres to soak for approximately 10 minutes. Pour off the water and then gently squeeze the fibres to remove excess water.
3. Add the damp fibres individually to the mordant solution using tweezers and gently stir the fibres briefly to ensure an even distribution of mordant into the fibres. Place the beaker on a hotplate and insert a thermometer into the solution.

Heat the mixture of fibres and mordant solution to 50°C–55°C over approximately 30 minutes and then leave at this temperature for 1 hour. Briefly and gently stir the fibres in the solution approximately every 15 minutes.

Note that wool and silk are more sensitive to sudden changes in temperature or vigorous agitation than cotton materials and so solutions containing the protein-based fibres should be heated slowly and stirred or swirled gently. Optional: placing a crystallising dish on the hotplate at the start of the experiment is an additional safety measure as it will contain spills and minimise clean-up in the unlikely event of the beaker breaking.

4. Switch off the heat and let the fibres cool to room temperature in the mordant solution.

5. Remove the fibres from the mordant solution using tweezers and transfer to a clean 250 cm³ beaker. Rinse the fibres thoroughly with tap water. The damp fibres are now ready for dyeing. Alternatively the fibres can be air dried and stored until required.

Note that the ratio of mordant to weight of fibres in this procedure is generous (1 g : approx. 3.5 g) compared to many standard recipes where the amounts of alum and aluminium acetate are 15% and 9% of the weight of fibres respectively. Alum is often the mordant of choice for protein (animal) fibres whilst aluminium acetate is widely used for the mordanting of cellulose (plant) materials. However, this is an area which the interested reader can explore further, for example, an investigation into the effectiveness of dyeing cellulose fibres with alum etc. The procedure above is a guide and the effect of mordanting at higher and lower temperatures and/ or for shorter or longer time periods can also be explored.

6.2.2   Dyeing Fibres with Madder, Weld and Cochineal

Materials

madder extract	0.10 g	not a hazardous substance
weld extract	0.10 g	not a hazardous substance
cochineal extract	0.01 g[a]	not a hazardous substance

deionised water *(if this is not available use rain or tap water)*

mordanted wool, silk and cotton fibres and fabrics

[a] *Accurately weighing small amounts (≤10 mg) of solids on a 2-decimal place balance is difficult and an alternative approach is to prepare and use a stock solution as follows. Weigh out a larger mass, e.g. 0.10 g, and dissolve this solid in a known volume of water. Then measure out a known volume of this stock solution and dilute accordingly to the required concentration.*

Equipment: hotplate, thermometer (0°C–100°C), 100 cm³ measuring cylinder, 100 cm³ beakers, spatulas, tweezers

---

Procedure
Although the madder, weld and cochineal dye extracts are not considered hazardous substances it is advisable to work in a well-ventilated area and wear gloves in addition to the standard PPE when carrying out these experiments.

1. Weigh out the dye extract into a 100 cm^3 beaker and add 50 cm^3 of water. Stir and then warm the mixture to approximately 50°C for 30 minutes. If all of the dye has not dissolved, filter the mixture to obtain a clear dyebath.
   Note that cochineal is a more concentrated dye than madder or weld and smaller amounts are used to give a reasonable colour.
2. Add one piece of each type of mordanted fibre (see Section 6.2.1 step 5) separately to the dyebath using tweezers and gently stir the mixture. If the mordanted fibres have been dried and stored they are first wetted as described in step 2 of procedure 6.2.1.
3. Place the beaker on a hotplate and insert a thermometer into the dyebath. Heat the dyebath containing the fibres to 50°C–55°C over approximately 30 minutes and then leave at this temperature for 1 hour. Briefly and gently stir the fibres in the solution approximately every 15 minutes.
4. Switch off the heat and let the fibres cool to room temperature in the dyebath.
5. Remove the fibres from the dyebath using tweezers and transfer to a clean 250 cm^3 beaker. Rinse the fibres thoroughly with tap water and air-dry.
6. Store your dyed and dried fibres in a resealable transparent plastic bag and attach a label.
   Note that the mass of dyestuffs above are only a guide and the shade can be varied by using greater or smaller quantities. The dyebaths can also be used again to dye more fibres but the colours will be paler. Other potential variations to the above procedure include changing the temperature and/ or length of time the fibres are in the dyebath or by adding a modifier. A modifier may change the colour of the dyed fibre and is added to the dyebath. The most common modifiers are acids (e.g. citric acid or acetic acid/ vinegar), bases (e.g. sodium carbonate) or metal salts (e.g. copper(II) sulphate or iron(II) sulphate). Only a few milligrams of a modifier are needed to effect a colour change on the fibres. Stock solutions of these compounds can be prepared and then an appropriate volume added to the dye bath.

6.2.3   DYEING WITH INDIGO*

Materials

indigo	0.15 g	**skin and eye irritant, possible respiratory irritant**
sodium carbonate (anhydrous)	0.20 g	**eye irritant**
thiourea dioxide	0.20 g	**health hazard: toxic and corrosive**
[$H_2N–C(=NH)–SO_2H$]		

deionised water *(if this is not available use rain or tap water)*
wool, silk and cotton fibres and fabrics
Equipment: hotplate, thermometer (0°C–100°C), 100 cm^3 measuring cylinder, 10 cm^3 measuring cylinder, 100 cm^3 beaker (×2), Pasteur pipette, spatulas, tweezers

---

Indigo is referred to as a vat dye as it is reduced to a soluble form in a reducing bath before being applied to the fibres.

Procedure

1. Weigh out 0.20 g of anhydrous sodium carbonate into a small vial and dissolve in 1 cm^3 of warm water. Cool the solution to room temperature, add 0.15 g of indigo and mix to a paste.
2. Heat 50 cm^3 of water in a 100 cm^3 beaker to 50°C–55°C on a hotplate.
3. Add the indigo/sodium carbonate mixture to the warm water. Rinse the indigo/sodium carbonate vial with several 1 cm^3 portions of water and add to the dye bath. Stir carefully and avoid creating bubbles of air.
4. Weigh out 0.20 g of thiourea dioxide into a small vial.
   CARE *This procedure must be carried out in a fume cupboard or in a well-ventilated area, preferably outside, using a face mask.*
5. *In the fume cupboard/well-ventilated area*, carefully add the thiourea dioxide to the aqueous indigo mixture and stir slowly and gently so as to avoid adding air to the solution. Cover the vat dye bath with aluminium foil.
6. Continue to heat the dye vat at 50°C–55°C for approximately 30 minutes or until the solution attains a clear green-yellow colour with a bronze surface. The green-yellow may be seen more clearly if a small amount of the solution is taken up into a pipette. If after 1 hour the dye vat is not a green-yellow colour add a further 0.10 g thiourea dioxide and wait for a few minutes. Repeat with further portions of thiourea dioxide if necessary.
7. Weigh the fibres in a 100 cm^3 beaker. Add approximately 50 cm^3 water and leave the fibres to soak for approximately 10 minutes. Pour off the water and then gently squeeze the fibres to remove excess water.
8. Immerse the squeezed fibres in the dye vat and immediately recover the container with the aluminium foil.
9. Leave the fibres in the dye vat for 10–15 minutes, keeping the temperature at 50°C–55°C.
10. Prepare a rinsing solution by heating another beaker of water to 50°C–55°C. This can be done whilst the vat is being prepared and the dye is being absorbed into the fibres.
11. Carefully remove the fibres from the dye vat and immerse them in the warm rinsing solution. When the fibres are removed from the vat they start to change colour from yellow to green to blue as the leucoindigo oxidises back to indigo. Leave the fibres in the rinsing solution for 20–30 minutes in the open air to complete the oxidation process.
12. Rinse the fibres well with tap water and then air-dry the samples. Place the dried dyed fibres in a resealable transparent plastic bag and attach a label.
    Step 8 can be repeated if a darker blue shade is required. Lighter shades can be obtained by reducing the immersion time in step 9. As with the dyebaths in Section 6.2.2 *the indigo vat dyebath can also be used again to dye more fibres. Carrying out steps 7 to 12 with fibres dyed as described in* Section 6.2.2 *is known as overdyeing and will generate different colours.*

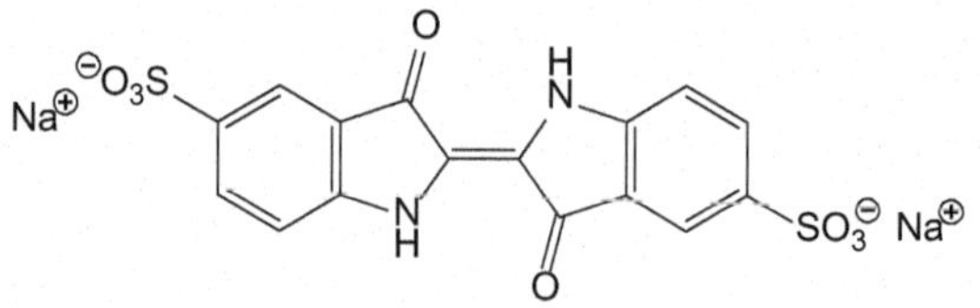

Indigo carmine

FIGURE 6.3 Structure of indigo carmine.

6.2.4 Dyeing with Indigo Carmine, a Semi-Synthetic Compound

Indigo carmine (Figure 6.3) is a blue synthetic dye prepared from indigo by an aromatic sulphonation process. The presence of the sulphonate $(-SO_3^-)$ groups on the molecule renders this compound water soluble. As the indigo carmine is an acidic organic salt it can form ionic bonds with basic groups in protein-based fabrics. Dyeing fabrics with indigo carmine is analogous to the procedures given in Section 6.2.2. It is recommended that concentrations of ~0.5–1.0 mg/cm^3 are used in the first instance as the dye is intensely coloured. The effects of dyeing mordanted and non-mordanted fibres with indigo carmine can be explored and this compound can also be used to overdye fibres already coloured with natural dyes. Indigo carmine may cause **allergic skin reactions** and so gloves are recommended when using this material.

6.3 GOLD NANOPARTICLES

Gold nanoparticles (AuNPs) can be formed by treating aqueous solutions of chloroauric acid, $HAuCl_4$ (also known as gold(III) chloride), with a reducing agent. Reduction of $HAuCl_4$ produces a colloidal suspension of nanometre-sized metallic gold(0) particles which are coloured due to the localised surface plasmon resonance effect. The observed colour of the colloid suspension depends on the size and shape of the AuNPs and ranges from red to blue and purple. The colour and optical properties of gold and other metallic nanoparticles have made these substances suitable for a broad range of applications including biosensors, medical diagnostics and dyes. It is this latter application which is the focus of the experiments in this section.

Reduction of gold(III) to gold(0) can be achieved using compounds including sodium borohydride ($NaBH_4$), organic acids such as ascorbic acid and citric acid and their salts, and some α-amino acids. Sodium ascorbate (vitamin C) is the chemical reducing agent used in two of the procedures below and the equation for the reaction is given in Figure 6.4. Longer reaction times and/or heating is required for the reduction of $HAuCl_4$ with citric acid and α-amino acids compared to the reduction with ascorbate. Formation of AuNPs from $HAuCl_4$ can also be effected by fruit peel extracts in a green synthesis approach and an article on this route is given in the *Further Reading* section at the end of this chapter.

Each of the three experiments below uses a 1 mM stock solution of $HAuCl_4$. Chloroauric acid is an expensive reagent and readers embarking on these experiments are advised to plan their work carefully. Read the procedures and watch the

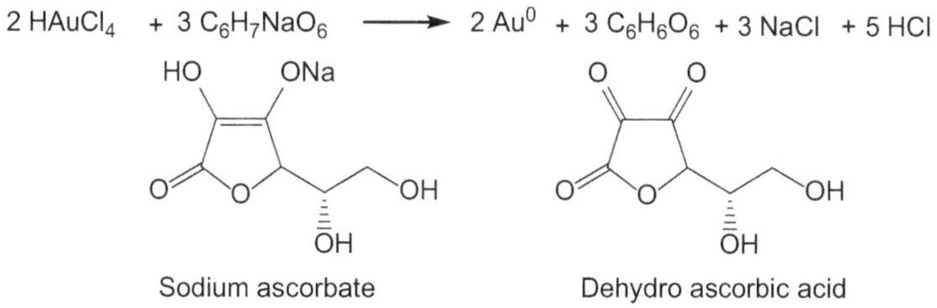

$$2 \text{ HAuCl}_4 + 3 \text{ C}_6\text{H}_7\text{NaO}_6 \longrightarrow 2 \text{ Au}^0 + 3 \text{ C}_6\text{H}_6\text{O}_6 + 3 \text{ NaCl} + 5 \text{ HCl}$$

Sodium ascorbate   Dehydro ascorbic acid

FIGURE 6.4 Reduction of chloroauric acid to elemental gold with sodium ascorbate.

videos of the experiments to decide what volume of 1 mM solution of HAuCl_4 is required for the planned work. For example, 4 mg of HAuCl_4 trihydrate (the purchased form of chloroauric acid used by the authors) is sufficient to give 10 cm^3 of a 1 mM solution. Small amounts of solid are best weighed out on a 3- or 4-decimal place balance for accuracy.

Procedures

Materials

chloroauric acid trihydrate, $\text{HAuCl}_4 \cdot 3\text{H}_2\text{O}$	a few mg	**health hazard: harmful if swallowed**
L-sodium ascorbate, $\text{C}_6\text{H}_7\text{NaO}_6$	0.100 g	not a hazardous substance
deionised water		

wool, silk and cotton fibres and fabrics

Equipment: hotplate fitted with a temperature probe or thermometer, sand bath, 100 cm^3 measuring cylinder, graduated 1 or 2 cm^3 pipette ($\times$2), small sample vials, spatulas, tweezers

---

Preparation of stock solutions

1. Prepare a 1 mM solution of chloroauric acid by dissolving the appropriate mass of solid $\text{HAuCl}_4 \cdot 3\text{H}_2\text{O}$ in the appropriate volume of water for the planned experiments. Transfer the resulting pale yellow solution to a dark coloured bottle.

   CARE *Weighing solid $HAuCl_4 \cdot 3H_2O$ must be carried out in a fume cupboard or in a well-ventilated area using a face mask and wearing gloves.*

2. Weigh sodium ascorbate (the sodium salt of ascorbic acid, 0.10 g) into a conical flask. Dissolve the solid in 50 cm^3 of water to give a 10 mM solution of sodium ascorbate. *Solutions of sodium ascorbate are susceptible to oxidation and can degrade over time and fresh solutions of this reagent should be prepared if the experiments are carried out over more than a few days.*

6.3.1 REDUCTION OF GOLD(III) IONS TO GOLD NANOPARTICLES WITH SODIUM ASCORBATE*

1. Using a graduated pipette, measure out 1.0 cm^3 of the 1 mM solution of HAuCl$_4$ and place in a small sample vial.
2. Add a few drops of the sodium ascorbate solution to the vial containing the gold(III) ions. Red coloured gold nanoparticles develop in a few seconds. Over time (several days) the nanoparticles aggregate and form a precipitate.

6.3.2 GOLD NANOPARTICLES AS A DYE: REDUCTION OF GOLD(III) IONS WITH SODIUM ASCORBATE*

In the procedure below a single piece of silk was dyed. The experiment can be repeated with the other types of fibre or the samples can be combined and dyed in the same reaction vessel. This will require larger volumes of the HAuCl$_4$ and sodium ascorbate solutions.

1. Using a graduated pipette, measure out 2.0 cm^3 of the 1 mM solution of HAuCl$_4$ and place in a small sample vial.
2. Add a small piece of silk (approx. 5 cm × 0.5 cm) to the vial containing the solution of gold(III) ions. The material can either be totally immersed in the solution or a portion of it dipped into the solution. Leave the material soaking in the solution for 10 minutes.
3. Using a graduated pipette, measure out 2.0 cm^3 of the sodium ascorbate solution and place in a small sample vial.
4. Using a pair of tweezers remove the silk from solution of gold(III) ions and transfer it to the sodium ascorbate solution. Leave the silk in the sodium ascorbate solution for 10 minutes or until the material has attained a suitable colour.
5. Using a pair of tweezers, remove the silk from the sodium ascorbate solution to a beaker of water, rinse well with water and then air-dry. Place the dried coloured piece of silk in a resealable transparent plastic bag and attach a label.
6. The residual HAuCl$_4$ solution may be reused and steps 2 to 5 repeated with further samples of silk. The colour of the dyed material becomes less intense with each dye cycle.

An alternative experiment, analogous to the natural dye procedure of Section 6.2.2, involves treating samples of the fibres with a pre-formed colloidal solution of gold nanoparticles prepared using procedure of Section 6.3.1.

6.3.3   Gold Nanoparticles as a Dye: A Protein in situ Redox Method*

In the procedure below a single piece of silk was dyed. Tyrosine residues of the protein silk fibres act as the reducing agent and the reaction requires heating for the synthesis of the AuNPs to occur. The procedure can also be carried out with other protein fibres such as wool or spider silk.

1. Using a graduated pipette, measure out 2.0 cm^3 of the 1 mM solution of HAuCl$_4$ and place in a small sample vial.
2. Add a small piece of silk to the vial containing the gold(III) ions. The material can either be totally immersed in the solution or a portion of it dipped into the solution.
3. Place the vial containing the reaction mixture in a sand bath on a hotplate fitted with a temperature probe or thermometer. Heat the sand bath to a temperature of 90°C and then incubate the reaction mixture at this temperature for 20–30 minutes or until the silk has attained a suitable colour.
4. Using a pair of tweezers remove the silk from the reaction vial and transfer it to a beaker of water. Rinse the silk well with water and then air-dry. Place the dried coloured silk in a resealable transparent plastic bag and attach a label.

Investigative experiments can also be carried out to determine which of the 20 proteinogenic α-amino acids are the most effective reducing agents in the synthesis of AuNPs. A fair testing approach would need to be adopted and appropriate volumes of the HAuCl$_4$, mass or moles of the α-amino acids and reaction temperature would all need to be considered when planning this type of experiment.

6.4   PHOTOGRAPHIC PROCESSES

The experimental procedures for the cyanotype process and the copper toning of a black-and-white print in this section both use potassium hexacyanoferrate(III) K$_3$[Fe(CN)$_6$] as one of the reagents. The cyanide groups in this reagent, and also those in the Fe$_4$[Fe(CN)$_6$]$_3$ (Prussian blue) and Cu$_2$[Fe(CN)$_6$] products from these photographic processes, are tightly bound to the Fe^{3+} or Fe^{2+} ions resulting in these compounds being stable and non-toxic. Any excess reagents and reaction mixtures can be washed down a sink with plenty of water. However, these compounds can release the highly toxic gas hydrogen cyanide if brought into contact with strong acids. Although it is unlikely in these experiments, care should be taken to ensure no strong acid is present in the sink or drain when disposing of materials.

Both the cyanotype and toning procedures are 'printing-out processes' as the exposed paper or print can be seen to change colour as the chemical reactions proceed. The light-sensitive paper used in the cyanotype process requires several minutes in bright sunlight to create an image, and it is possible to work with this photographic material in a room with windows with the lights switched off. The toning experiments can be carried out under normal room lighting conditions.

6.4.1 CYANOTYPE PHOTOGRAPHY

Cyanotype printing was developed in the early 1840s by John Herschel and it creates a white image on a deep blue background. The procedure involves coating paper with a light-sensitive solution of ammonium iron(III) citrate, $(NH_4)Fe(C_6H_4O_7)$, and potassium hexacyanoferrate(III), $K_3[Fe(CN)_6]$.[1] After drying, the paper is exposed to UV light or sunlight for a few minutes. During the exposure, the Fe^{3+} in ammonium iron(III) citrate is reduced to the Fe^{2+} state and the citrate is oxidised to 3-oxopentanedioic acid (Figure 6.5). The reduced Fe^{2+} ions then react with hexacyanoferrate(III) ions to give the blue $Fe^{III}[Fe^{II}(CN)_6]^-$ complex. The charge is balanced by K^+ ions in 'soluble' Prussian blue $KFe^{III}[Fe^{II}(CN)_6]$ or with Fe^{3+} ions in iron(III) hexacyanoferrate(II), $Fe^{III}_4[Fe^{II}(CN)_6]_3$, the compound commonly known as Prussian blue which is insoluble in water. Objects placed on the sensitised paper will block the sunlight and the chemical reactions which take place in the cyanotype process will not occur in these areas. After sufficient exposure, the paper is washed in water which removes the soluble unexposed reactants leaving a white image of the object on a blue background.

Semi-transparent objects such as plastic rulers will create negative images on the paper whereas opaque objects such as keys or botanical specimens (leaves, ferns etc.) give silhouettes of their outlines which are known as photograms. Printing black-and-white negatives using this technique will produce a positive image on the sensitised paper.

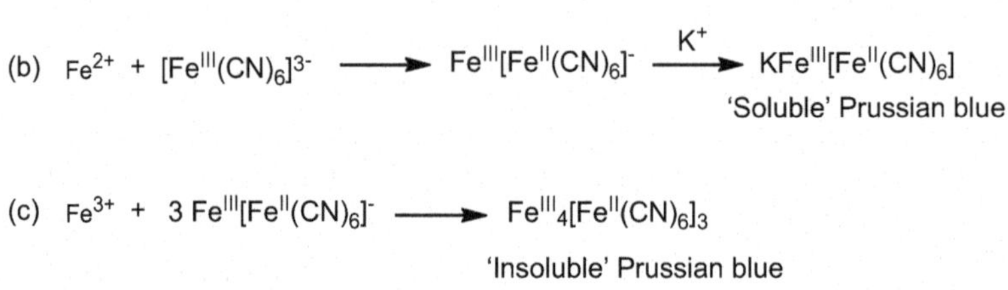

FIGURE 6.5 Equations for the reactions occurring in cyanotype printing. (a) In sunlight, citrate reduces Fe^{3+} to Fe^{2+} and is itself oxidised to 3-oxopentanedioic acid. (b) Electron transfer from Fe^{2+} to the hexacyanoferrate(III) complex generates a mixed valence complex which can form the compound 'soluble' Prussian blue. Despite the name, this form of the blue pigment is however insoluble but the particles can form a finely dispersed suspension which makes it appear to be soluble in water. (c) Formation of the more insoluble form of Prussian blue. Note that the oxidation state of iron ions in complexes and coordination compounds is indicated by Roman numerals whilst Arabic numerals show the charge of full ions.

Materials

| potassium hexacyanoferrate(III), | 1.0 g | **eye irritant** |
| $K_3[Fe(CN)_6]$ | | |

| ammonium iron(III) citrate | 2.5 g | not a hazardous substance |

deionised water

A4 or artist's watercolour paper (the above quantities will coat 5–6 sheets of paper)

Equipment: 25 cm³ conical flask or small beaker (×2),
10 cm³ measuring cylinder, glass stirring rod, small plastic container,
brush or sponge, spatulas, a drying line (string and pegs), timer, plastic
tray (for washing print in the sink)

Cyanotype is known as a contact printing process as the print is the same size as the original material. It became widely used in the commercial reproduction of architectural and engineering drawings from the 1870s to 1960s when it was replaced by a diazo-based procedure and later by large-format photocopiers. The original technical drawings were traced onto tracing paper or clear plastic using black ink and these were laid onto the sensitised paper and then exposed to sunlight. The clear background area became dark blue and the black line drawings reproduced as a white line. Artistic readers carrying out the cyanotype printing experiments can also create their own drawings using black marker pens and transparent film (or tracing paper) which can then be used to produce blueprints. Blueprints can also be created from digital black-and-white photographs taken with a phone or digital camera. These photographs first need to be converted, using the appropriate software, into negative images which are then printed or photocopied onto transparency film.

Procedure

Preparation of the sensitised paper
CARE Drips and spillages of the light-sensitive solution must be cleared up immediately as they will develop into permanent blue stains on surfaces or clothing upon exposure to light.

1. Weigh 1.0 g of potassium hexacyanoferrate(III) into a small conical flask (or beaker) and add 10 cm³ of distilled water. Stir until dissolved to give an orange solution.
2. Weigh 2.5 g of ammonium iron(III) citrate into a conical flask (or beaker) and add 10 cm³ of distilled water. Stir until dissolved to give a green solution.

   When the two solutions are mixed it will become light sensitive and so steps 3 to 5 must be carried out in a darkened room. The room does not have to be absolutely dark but switch off lights and close curtains/blinds.
3. Mix the two solutions together and pour the resulting (light-sensitive) mixture into a small plastic container – this will act as your paint pot.

4. Place a piece of A4 paper on newspaper and, using the brush or sponge, coat the paper with a layer of the solution from step 3. Leave a margin around the area coated to aid handling of the paper. The paper will turn a pale green colour. Let the paper drip over the newspaper to remove excess solution and then peg it onto the drying line, again under subdued lighting conditions. Repeat this step with fresh pieces of paper until all the solution has been used up. *You should be able to coat ~5 or 6 sheets with this quantity of solution.*

5. Once the papers are dry place them inside a cardboard box and then keep the box in a cupboard or black plastic bag.
     Note that the paper must be completely dry before making the prints as any moisture left in the paper will act as a developer and the clarity and detail of the print will be compromised.

Making cyanotype prints

1. **This step should be done in a darkened room.**
     Remove a sheet of the sensitised paper from the storage box and place it in a cardboard box or cover with a black cloth. This prevents exposure in transit when the paper is being moved from indoors to the sunlight outside. **The sensitiser chemicals on the paper may damage some objects so it may be prudent to place the paper in a plastic wallet for the printing of precious materials.**

2. Select the object that you wish to print, e.g. key, ruler, leaves, stencils etc. but ideally an item that is relatively flat. Remove the paper from the transit box, place the objects for printing onto the sensitised paper and expose the paper to sunlight. Outside on a sunny day is perfect but a window sill is an adequate alternative. The paper exposed to the sun will gradually go blue and the parts covered by the object will remain green.
     You may wish to make a test using a strip of paper first to establish the correct exposure time. To do this place a sheet of cardboard over your paper+object and then carefully move the carboard over the paper to enable different parts of the sheet to receive the sunlight for different lengths of time, e.g. 15 seconds, 1 minute, 2 minutes, 5 minutes. Repeat until you have optimised the exposure time. Longer exposure times will be necessary if it is a cloudy or overcast day. Cyanotype printing in the late autumnal or winter months may require exposure times of hours rather than minutes. Black-and-white negatives or botanical specimens may need to be held in place during the exposure using a sheet of glass and this will also give a sharper image. Place the negative or botanical material on the glass in step 1 above, followed by the sensitised paper and then a sheet of cardboard to give rigidity. Clip the layers together; a frameless picture frame is perfect for this procedure. The glass will absorb sunlight so you will have to lengthen your exposure time – again experiment to optimise.

3. Once you have exposed the paper to the sun for an appropriate length of time take the paper to the sink indoors for washing.

4. Wash the paper in water to remove the (soluble) green chemical mixture; the (insoluble) blue colour will remain. *Use running tap water for 5–10 minutes or until the highlights have cleared to white and there is no longer any yellow-green colouration in the water.*

5. Hang up your cyanotype print on the line to dry. *There is some variability in the structure of Prussian blue and the blue colour may darken as the paper dries in the air as a result of oxidation processes.*

6. Write notes on the procedure in your lab notebook and it is often useful to record information about the conditions of the day and exposure times etc. on the back of your dried cyanotype print.

The blue colour of the cyanotype print can subsequently be altered through toning experiments and information on this area can be found through the material listed at the end of this chapter. Another avenue which readers may be interested in exploring is the creation of cyanotype prints on materials other than paper; cotton or linen fabrics, wood or other materials which can absorb the sensitiser solution and withstand the rigours of wet processing can be subjected to cyanotype processing.

6.4.1.1 Prussian blue beyond the cyanotype

The colour and insolubility of Prussian blue has given this compound a rich history as a pigment. The formula, $Fe^{III}_4[Fe^{II}(CN)_6]_3$, of Prussian blue shows that this is a mixed valence coordination compound and the colour arises from the transfer of an electron between the Fe(II) and Fe(III) ions. This process occurs when light at ~680 nm in the orange-red region of the visible spectrum is absorbed and the reflected light therefore appears blue as a result. The Fe(II) and Fe(III) ions alternate in the structure and are linked by a CN group with the carbon atom coordinated to the Fe(II) ion. The Fe(III)–NC–Fe(II) sequence repeats in three dimensions to form a highly porous and stable polymeric structure which can take up and trap molecules or ions, up to a radius of 182 pm, in the lattice. In medicine this feature has been exploited to develop treatment protocols for radioactive caesium and thallium poisoning; toxic $^{137}Cs^+$ and Tl^+ ions are trapped in the Prussian blue lattice which can then be safely excreted by the body.

The insoluble $Fe^{III}_4[Fe^{II}(CN)_6]_3$ and soluble $KFe^{III}[Fe^{II}(CN)_6]$ forms of Prussian blue have similar lattice structures and in the latter compound, the K^+ ions occupy sites in the open channel framework structure. Prussian blue can be oxidised to the all Fe(III) compound Prussian yellow, $Fe^{III}[Fe^{III}(CN)_6]$, or reduced to Prussian white, $K_2Fe^{II}[Fe^{II}(CN)_6]$, which contains iron in only the +2 oxidation state. Prussian white slowly reoxidises in air, or more rapidly by treatment with dilute hydrogen peroxide (H_2O_2), and the blue colour returns. The electron transfers in these redox processes take place without changing the essential lattice structure of Prussian blue and the change in charge is balanced by the flow of K^+ ions in or out of the lattice. A channel structure which can reversibly accommodate alkali metal ions combined with a corresponding redox reaction of a transition metal ion are features of ion storage cells. Many research groups are synthesising Prussian blue analogues with the general formula $AM_1[M_2(CN)_6]$ (A = alkali metal ion, M_1, M_2 = transition metal ions) which have a similar framework structure to Prussian blue and are exploring the potential of these compounds as sustainable cathode material for rechargeable batteries.

6.4.2 COPPER TONING OF BLACK-AND-WHITE PHOTOGRAPHS

Toning of black-and-white photographs is a way of adding interesting colours into a print for aesthetic reasons. The toner chemically changes the metallic silver in the emulsion. Copper toning gives a warm reddish brown to a photograph as the silver in the image is replaced by an insoluble copper compound in a two-step process.

1. Hexacyanoferrate(III) ions $[Fe(CN)_6]^{3-}$ in the toner solution oxidise metallic silver in the print (which is the black in the image) and is reduced to the hexacyanoferrate(II) complex $Fe(CN)_6^{4-}$:

$$Ag\,(s) + Fe(CN)_6^{3-}\,(aq) \rightarrow Ag^+\,(aq) + Fe(CN)_6^{4-}\,(aq)$$

2. The free Cu^{2+} ions in the toner solution react with the $Fe(CN)_6^{4-}$ complex to form a precipitate of copper(II) hexacyanoferrate(II), $Cu_2Fe(CN)_6$:

$$2\,Cu^{2+}\,(aq) + Fe(CN)_6^{4-}\,(aq) \rightarrow Cu_2Fe(CN)_6\,(s)$$

The precipitate is formed at the location of the original black silver image and it is this precipitate which is responsible for the colour of the toned print. Where the original image is dark, a large amount of $Cu_2Fe(CN)_6$ is formed and the new colour is intense.

The silver within the black-and-white photograph can also be replaced by other metallic compounds, for example, silver sulphide (Ag_2S) or silver selenide (Ag_2Se). Silver sulphide gives sepia photographs their characteristic brown colour whilst silver images become reddish in selenium toning.

Materials

copper(II) sulphate pentahydrate, $CuSO_4 \cdot 5H_2O$	1.30 g	**harmful to eyes and if swallowed**
potassium hexacyanoferrate(III), $K_3[Fe(CN)_6]$	1.10 g	**eye irritant**
citric acid monohydrate, $C_6H_8O_7 \cdot H_2O$	6.86 g	**eye irritant, possible respiratory irritant**
potassium carbonate, K_2CO_3	6.74 g	**skin and eye irritant, possible respiratory irritant**

deionised water

black-and-white photographs; above quantities are sufficient to treat two 5 × 7 inch prints

Note: if old B&W prints are not available, then photographs for this experiment can be obtained using a process-paid disposable B&W camera with e.g. Ilford HP5 film.

Equipment: plastic tray which is a suitable size to accommodate the photographic print and the toning solution, 250 cm^3 beaker (×2), 100 cm^3 measuring cylinder, weighing boats or glass sample vials, glass rod, plastic or wooden tongs, spatulas, timer (optional), drying line (string and pegs)

Procedure

The toning solution is formed by mixing equal volumes of two stock solutions:

Solution A: Copper sulphate ($CuSO_4 \cdot 5H_2O$, 1.3 g) and potassium citrate ($K_3C_6H_5O_7$, 5.0 g) dissolved in 200 cm^3 deionised water.

Solution B: Potassium hexacyanoferrate(III) [$K_3Fe(CN)_6$, 1.1 g] and potassium citrate ($K_3C_6H_5O_7$, 5.0 g) dissolved in 200 cm^3 distilled water.

The potassium citrate for the stock solutions is prepared from the neutralisation reaction between potassium carbonate and citric acid (Figure 6.6).

Preparation of Solution A

1. Weigh out citric acid monohydrate 3.43 g into a 250 cm^3 beaker. Measure out 100 cm^3 deionised water into a measuring cylinder, add this volume to the citric acid in the beaker and stir the mixture until all the solid has dissolved.
2. Weigh out 3.37 g potassium carbonate into a small beaker and carefully add the solid, in small portions, with stirring to the solution of citric acid. Effervescence occurs due to the formation of CO_2 in the neutralisation reaction. Ensure the effervescence has largely subsided between the addition of each portion of potassium carbonate. Measure out a further 100 cm^3 deionised water and add to the potassium citrate solution in the 250 cm^3 beaker. Thoroughly stir the 200 cm^3 solution of potassium citrate.
3. Weigh out copper sulphate pentahydrate (1.30 g) into a small sample vial or weighing boat and add the solid to the solution of potassium citrate. Stir the mixture with a glass rod to dissolve the solid into the solution. Label this as Solution A.

Preparation of Solution B

Repeat steps 1 and 2 above to prepare a second 200 cm^3 solution of potassium citrate. Weigh out potassium hexacyanoferrate(III) (1.10 g) into a small sample vial or

$$3\ K_2CO_3 + 2\ C_6H_8O_7 \cdot H_2O \longrightarrow 2\ K_3C_6H_5O_7 + 5\ H_2O + 3\ CO_2$$

M_r (g/mol)   138.2   210.0   306.3   18.0   44.0

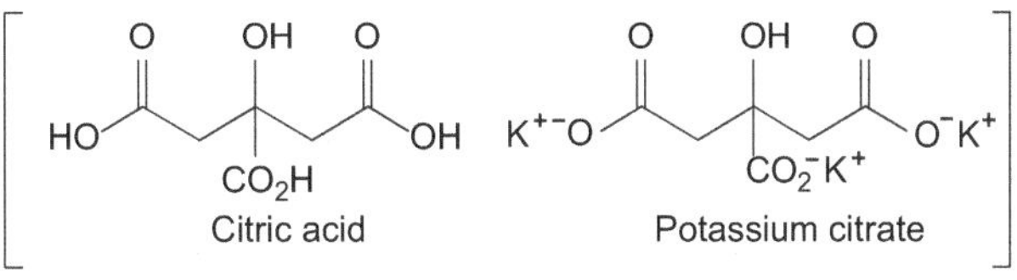

FIGURE 6.6 Reaction between potassium carbonate and citric acid to form potassium citrate. Formation of 5.0 g (0.0163 mol) of potassium citrate requires a 3:2 molar ratio of potassium carbonate (3.37 g, 0.0244 mol) and citric acid monohydrate (3.43 g, 0.0163 mol).

weighing boat and add the solid to the solution of potassium citrate. Stir the mixture with a glass rod to dissolve the solid into the solution. Label this as <u>Solution B</u>.

<u>Toning a photograph</u>

1. Select a plastic tray, e.g. ice-cream tub, that is of the appropriate size to accommodate the black-and-white print.
2. Equal volumes of <u>Solution A</u> and <u>Solution B</u> are required for the toning solution. Work out the volume of toning solution needed to treat the photograph based on filling the plastic tray to a depth of approximately 1 cm.
3. Measure out the required volumes of <u>Solution A</u> and <u>Solution B</u> based on the calculation in step 2.
4. Transfer the measured volume of <u>Solution A</u> to the plastic tray. Add the same volume of <u>Solution B</u> to the tray containing <u>Solution A</u>. Gently mix the toning solution with a glass stirring rod or by gently rocking the tray from side to side.
5. In a separate tray, soak a black-and-white photograph in deionised water for about a minute to ensure the emulsion is fully wet.
6. Remove the black-and-white print from the deionised water bath and place in the toning solution. Gently rock the tray to send the solution over the print. The shade of copper tone is adjusted by the length of time the print is in the toning solution; the longer it is in the solution the more copper-coloured it becomes.
7. When the photographic print has the desired copper shade, transfer it (using the tongs) to the bath of deionised water to remove most of the excess toning solution and then wash the print under the cold tap.
8. Hang up the toned print to dry.

Use fresh toning solutions when treating subsequent photographs. Note that the role of the citrate is to act as a chelating agent in order to 'trap' the copper ions such that the concentration of free Cu^{2+} ions in solution is much lower and does not form a precipitate with the ferricyanide ions.

6.5 SYNTHESIS OF TRIS(8-HYDROXYQUINOLINE) ALUMINIUM(III), ALQ$_3$, A LUMINESCENT MATERIAL*

Alq$_3$ is an organic electron transport and luminescent material widely used in OLEDs. In OLEDs, layers of semiconducting organic material, often incorporating Alq$_3$, are sandwiched between two electrodes. A current of electrons flowing through the device removes electrons from the valence band, creating positive 'holes' and inserts the electrons into the higher energy conduction band of the organic layer. Electrostatic forces bring a negative electron and a positive hole towards each other to form a bound (paired) state called an exciton. Excitons have a very short lifetime and when the electrons and holes recombine, energy is released in the form of light. Electron-hole recombination in Alq$_3$ results in the emission of bright green light.

Irradiation of Alq_3 with ultraviolet light can promote an electron into a higher energy orbital. When this excited state decays it is accompanied by the re-emission of light, but of a longer wavelength, and Alq_3 is seen to glow with green coloured light. Emission of lower energy light by a substance following absorption of higher energy radiation is known as luminescence and if the luminescence ceases as soon as the source of excitation is removed it is called fluorescence. If the emission of light persists the effect is known as phosphorescence. The aim of this experiment is to synthesise and observe the fluorescence of Alq_3.

The Alq_3 molecule consists of three hydroxyquinoline ligands bonded to one aluminium ion (Al^{3+}). In the procedure below Alq_3 is prepared by the reaction of 3:1 molar ratio of 8-hydroxyquinoline with aluminium acetate [$Al(OCOCH_3)_3$] in an aqueous ethanol solution (Figure 6.7). Alq_3 crystallises from the reaction mixture as a yellow solid and is collected by suction filtration.

There is intense interest in other metal-chelate systems in the search for new opto-electronic materials. For example, one of the authors and his research group have synthesised new light-emitting zinc chelates of fluorinated benzothiazole derivatives that are structurally related to oxyluciferin, the compound responsible for light production in fireflies and other bioluminescent insects.

Materials

8-hydroxyquinoline (C_9H_7NO)	290 mg	**health hazard: toxic if swallowed, serious eye damage, may cause allergic skin reaction**
aluminium acetate [$Al(OCOCH_3)_3$] (also known as aluminium triacetate)	135 mg	**skin and eye irritant**
ethanol		**flammable, irritant to eyes**
deionised water		

Equipment: magnetic stirrer/hotplate (ideally fitted with a temperature probe), aluminium heating block (for 25 cm³ round bottomed flask), 25 cm³ round bottomed flask, magnetic stirrer bar, air condenser, 10 cm³ measuring cylinder, Pasteur pipette, suction filtration funnel and flask, membrane pump or water aspirator, UV 254/365 nm wavelength lamp

---

Procedure

1. Weigh out 8-hydroxyquinoline (290 mg 2.00 mmol) into a small sample vial. Transfer the solid to a 25 cm³ round bottomed flask.
   CARE *This step must be carried out in a fume cupboard.*
2. Using a 10 cm³ measuring cylinder, measure out ethanol (8 cm³) and add this solvent to the flask containing the 8-hydroxyquinoline. Add a small magnetic stirrer bar to the reaction flask.
3. Transfer the reaction flask to an aluminium heating block located on a hotplate stirrer *in the fume cupboard*. Warm and stir the solution to dissolve the 8-hydroxyquinoline.
4. Fit an air condenser to the reaction flask and heat the 8-hydroxyquinoline until it boils.

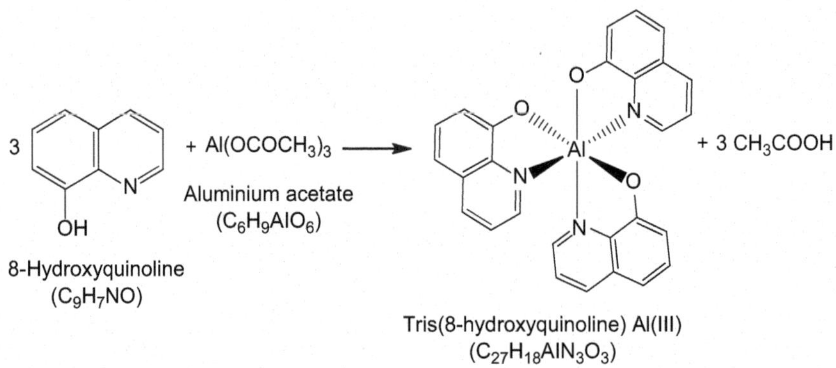

FIGURE 6.7   Synthesis of tris(8-hydroxyquinoline) aluminium(III). In the synthesis three H$^+$ ions are lost from the OH groups as the Al^{3+} ion coordinates.

5. Weigh out aluminium acetate (135 mg, 0.66 mmol) into a sample vial and dissolve the solid in water (2 cm^3).
6. Using a Pasteur pipette, add the aluminium acetate solution dropwise down through the air condenser to the refluxing solution of 8-hydroxyquinoline.
7. Reflux the reaction mixture for 30 minutes. Crystals start to precipitate from the solution after several minutes.
8. Switch the heat off, raise the reaction flask from the heating bock and leave the mixture to cool to room temperature.
9. Transfer the reaction flask to a beaker of iced water for approx. 30 minutes or place the reaction flask in the fridge. Also cool test tubes of water and ethanol in iced water/fridge.
10. Remove the magnetic stirrer bar and then collect the product by suction filtration. Rinse the reaction flask with several portions of iced water to ensure all the product is transferred to the sinter or small Büchner funnel.
11. Wash the collected solid product in the sinter/Büchner funnel with cold ethanol (2 × approx. 2 cm^3).
12. The product is dried by either leaving the solid in the sinter funnel and continuing with the vacuum suction for several minutes and/or spreading the solid on filter paper and allowing it to dry in the air.
13. Transfer the tris(8-hydroxyquinoline) aluminium(III) crystals to a weighed and labelled sample vial.
14. Weigh the vial containing the product and calculate the percentage yield.
      Ideally the product should be dried thoroughly in a desiccator under vacuum before it is weighed and analysed. Expected yield is approximately 120 mg.
15. Transfer the sample to a small watch glass and view the compound under UV light using a long wavelength (365 nm) setting.

CARE *UV light is particularly hazardous and can cause serious damage to eyes and skin. Never look directly at the light source and it is recommended that gloves are worn. Ideally, the compound should be observed in a UV viewing cabinet.*

The reaction can also be carried out between 8-hydroxyquinoline and zinc(II) acetate using a 2:1 molar ratio of the starting materials, for example, 290 mg (2.0 mmol) of 8-hydroxyquinoline and 220 mg (1.0 mmol) of zinc acetate dihydrate [Zn(OCOCH₃)₂·2H₂O]. The product from this reaction is bis(8-hydroxyquinoline) zinc(II). This compound is less soluble in ethanol than Alq₃ and a higher isolated yield is obtained.

6.6 CHOLESTERYL BENZOATE, A LIQUID CRYSTAL

Cholesteryl benzoate was the first in a class of compounds now known as liquid crystals. Liquid crystals have a two-stage melting point with an opaque liquid (nematic) phase forming between the solid compound and the clear liquid state. The transition between different phases corresponds to a change in the three-dimensional ordering of the molecules. Chiral liquid crystal molecules such as cholesteryl benzoate spontaneously align into a spiral helical structure in the opaque (chiral nematic) phase (see Chapter 5 Figure 5.9).

In some chiral compounds, such as cholesteryl benzoate, a helical structure with a short pitch can develop into a three-dimensional fluid lattice of twistedness. The molecules adopt helical ordering along two perpendicular axes as opposed to the single twist of an ordinary chiral nematic helix and the double-twist cylinders are packed into cubic lattices. This highly ordered structure has lattice parameters which are comparable to the wavelengths of visible light and selective Bragg reflections can occur resulting in bright blue-violet colours. This phase, which is called the 'blue phase', exists in a narrow temperature range (one or two degrees) between the chiral nematic and isotropic phases. Blue phases can be considered as three-dimensional self-assembled optical nanostructures and may have a role not only in future generations of electro-optic devices but also in biomedical applications and other areas.

The aim of this experiment is to synthesis cholesteryl benzoate, determine the two transition temperatures and observe the 'blue phase'. Cholesteryl benzoate is readily synthesised from cholesterol and benzoyl chloride (an activated form of benzoic acid) (Figure 6.8) and is purified by recrystallisation. The purified product is analysed by thin layer chromatography (TLC) and spectroscopic techniques. Infrared (IR) and proton nuclear magnetic resonance (¹H NMR) spectra on the product prepared by the authors are included and key features of the data are highlighted. The liquid crystal transition temperatures of cholesteryl benzoate are determined using a simple melting point instrument whereupon the blue phase is observed.

6.6.1 Synthesis of Cholesteryl Benzoate*

Materials

cholesterol	0.97 g	not a hazardous substance
pyridine (anhydrous)	3 cm³	**flammable, harmful if swallowed, irritant to skin and eyes**
benzoyl chloride (density 1.21 g/cm³)	0.87 cm³	**health hazard: toxic if inhaled, harmful if swallowed, causes severe skin burns and eye damage**

calcium chloride   **irritant to eyes**
  (anhydrous)
ethyl acetate   **flammable, irritant to eyes**
deionised water

Equipment: magnetic stirrer/hotplate (ideally fitted with a temperature probe), aluminium
  heating block (for 25 cm³ round bottomed flask), 25 cm³ round bottomed flask, magnetic
  stirrer bar, air condenser, drying tube, nitrogen balloon and needle, 5 cm³ syringe and
  needle, 1 cm³ volumetric pipette, suction filtration funnel, flask and pump

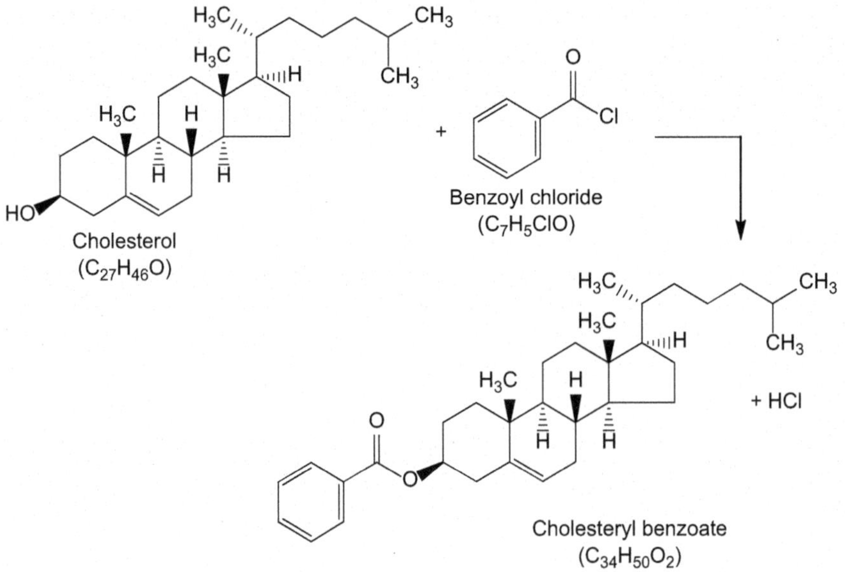

FIGURE 6.8 Synthesis of cholesteryl benzoate.

Procedure

1. Weigh out 0.97 g (2.5 mmol) of cholesterol into a 25 cm³ round bottomed
   flask and add a small magnetic stirrer.
2. Add anhydrous pyridine (3 cm³) to the cholesterol and place a calcium chloride drying tube onto the reaction flask. Transfer the reaction flask to an aluminium heating block set on a hotplate stirrer and then warm the mixture to
   dissolve the cholesterol in the pyridine.
   CARE *This procedure must be carried out in a fume cupboard.*
   *Anhydrous or dry pyridine is the solvent for the reaction and such sol-
   vents are contained in bottles fitted with a rubber seal. Such solvents are
   dispensed and transferred using a syringe technique. A balloon of nitrogen
   is fitted with a needle which is inserted through the rubber seal into the dry
   solvent bottle. This is to maintain an inert atmosphere in the bottle during
   and after withdrawal of the dry solvent. A Luer lock syringe (5 cm³) fitted*

with a long needle is inserted through the rubber seal and into the solvent. The plunger is slowly pulled back to allow the syringe to fill with solvent to the required volume (3 cm³). A vacuum is created in the bottle as the solvent is removed and the volume removed is replaced by the inert gas which is pulled into the bottle from the balloon. The syringe is then withdrawn from the solvent bottle and the solvent is quickly transferred to the reaction flask containing the cholesterol and then the drying tube, which is temporarily removed, is replaced.

3. Using a graduated volumetric pipette, measure out 0.87 cm³ (7.5 mmol, 3 equivalents) of benzoyl chloride and add this volume directly to the reaction flask. The drying tube is temporarily removed from the reaction flask during the addition of the reagent.

   CARE *This procedure must be carried out in a fume cupboard.*

4. Add an air condenser to the reaction flask and reposition the drying tube to the top of the condenser.

5. Heat and stir the reaction mixture until the solvent starts to boil and then maintain this refluxing state for 15 minutes.

6. Switch off the heat, raise the reaction flask from the heating bock and allow the solution to cool to room temperature. During cooling a solid forms in the reaction flask.

7. Add approx. 10 cm³ of water to the reaction flask. Stir the mixture to break up the solid product.

8. Remove the magnetic stirrer bar and then collect the product by suction filtration. Rinse the reaction flask with several portions of water to ensure all the product is transferred to the sinter or small Büchner funnel. Wash the solid on the sinter funnel with water.

9. Switch off the vacuum. Add approx. 5 cm³ of ice-cold methanol to the solid in the sinter funnel and stir the mixture thoroughly. Apply the vacuum suction to remove the methanol. Repeat this procedure with a further portion of cold methanol. *Optional: use a clean receiving flask for this step.*

10. Tip the product onto a piece of filter paper and then transfer the solid to a small (25 cm³) conical flask for recrystallisation.

11. Add approx. 6–8 cm³ of ethyl acetate (ethyl ethanoate) to the crude product and place a small magnetic stirrer bar into the conical flask.

12. Heat the mixture on a hotplate stirrer until the solvent boils. If the compound does not completely dissolve in this volume of solvent add further portions of ethyl acetate until complete dissolution occurs.

13. Remove the flask from the heat, remove the magnetic stirrer bar and allow the solution to cool to room temperature. The product should crystallise readily. Cool the flask in a beaker of iced water or place it in the fridge for approx. 30 minutes. Also cool a test tube of ethyl acetate in iced water/fridge.

14. Collect the crystals by suction filtration. Rinse the flask with several portions of ice-cold ethyl acetate to ensure all the recrystallised product is transferred to the sinter/small Büchner funnel.

15. Rinse the collected pure product on the sinter/Büchner funnel with a little more cold ethyl acetate ($2 \times$ approx. 2 cm^3).

16. Dry the crystals by either leaving them in the sinter funnel and continuing with the vacuum suction for several minutes and/or spreading the crystals on filter paper and allowing them to dry in the air.

17. Transfer the pure cholesteryl benzoate to a weighed and labelled sample vial.

18. Weigh the vial containing the product and calculate the percentage yield.

Ideally the product should be dried thoroughly in a desiccator under vacuum before it is weighed and analysed. Expected yield is approximately 1.0 g.

6.6.2 ANALYSIS OF CHOLESTERYL BENZOATE

Three techniques of analysing the cholesteryl benzoate product are described below. The first two, melting point determination and TLC, are techniques which are carried out routinely in school laboratories whilst the third involves recording IR and/or ^{1}H NMR spectra which requires access to instrumentation found in university chemistry or commercial research laboratories. Whilst the reader would find it rewarding to obtain and evaluate such spectra on their own synthetic sample, the authors appreciate that this may not be possible. The authors have therefore provided the spectra obtained on the sample prepared using the procedure described in Section 6.6.1.

6.6.2.1 Melting Point Determination

Measuring the melting point is one of the oldest analytical techniques used to indicate the purity of a compound. Impurities present in a substance create defects in the lattice structure and cause the melting point to be lower than that of the pure compound. Melting point apparatus range in complexity from an oil bath containing a thermometer which is heated with a Bunsen burner to a microscope with a heated stage. The authors recommend a capillary tube method using an electrically heated block apparatus. This type of apparatus has an integrated light source, the rate of heating can be easily controlled and the compound is viewed through a magnifying lens.

Procedure

1. Place a sample of the recrystallised cholesteryl benzoate material in a capillary tube closed at one end. Use capillary tubes designed for melting point determination as these are slightly wider than those used in the TLC technique [see method (b) below] and are consequently easier to fill.

   Press the open end of the capillary tube into a small pile of the (ideally powdered) compound on a watch glass or filter paper. This causes a small plug of the compound to become trapped in the mouth of the tube. The trapped solid is moved to the sealed end by gently tapping the sealed end of the capillary tube on the bench. Repeat this procedure until the capillary tube is packed with solid to a depth of about 1 cm. Note that this is a larger amount of compound than is normally used in a melting point determination but this mass will enable the transitions between solid to opaque liquid to transparent liquid to be observed more clearly.

2. Place the capillary tube in the melting point apparatus. Use the heating control to give an initial temperature rise of about 10°C/minute. When the temperature is about 20°C below the expected melting point temperature, reduce the rate of heating to about 2°C/minute.

    In the literature cholesteryl benzoate melts from a solid to a visibly cloudy liquid crystal by a temperature of 150°C and clears to an isotropic (clear) liquid by 180°C.

3. Record the melting point range at each of the expected transitions. At the lower transition this is the temperature from the appearance of the first drop of liquid to the last crystal of solid disappearing. Watch for the blue phase as the cholesteryl benzoate melts to a clear liquid and note the temperature range over which it occurs. Also record the temperature when the opaque liquid completely clears.

4. Switch off the heat and observe what happens when the sample cools.

    A blue-violet colouration may also be observed when cholesteryl benzoate transitions between the crystalline and the chiral nematic (opaque) phase.

6.6.2.2   Thin Layer Chromatography

TLC is an analytical technique which can identify how many components there are in a sample and thus determine its purity. It is one of a suite of chromatographic techniques that uses the interplay between the polarity of molecules with two other substances, called the stationary and mobile phases, to separate the different components in a mixture. The stationary phase is a porous finely divided solid which adsorbs molecules through intermolecular interactions and the mobile phase is a solvent which flows through the stationary phase. As the solvent rises up the plate through capillary action there is constant competition for molecules to dissolve in the solvent, and be carried up the plate, and to interact with the stationary phase. Compounds which interact strongly with the stationary phase will be held back whilst those which are less tightly adsorbed and are more soluble in the mobile phase will travel further up the plate.

The stationary phase is frequently a thin layer of silica on a solid plastic backing and the solvent tank for the TLC plates is a jar with a lid or a beaker covered with a watch glass. When the mobile phase nears the top of the TLC plate the plate is removed from the solvent tank and dried. In this experiment both the compounds are colourless and the positions of the spots on the TLC plate are visualised using two techniques. First, the plate is viewed under UV light to reveal compounds which have a chromophore such as an aromatic ring. Secondly, the TLC plate is treated with a reagent that reacts with the compounds to form a coloured product. Identifying how many spots a sample produces on the TLC plate will indicate the purity of the material; a pure compound will produce a single spot on the TLC plate whilst a reaction that has not gone to completion will show spots from both the starting material and the product. More information on this technique can be found in the *Further Reading* section.

Materials

iodine crystals (few mg, as a TLC visualisation agent)	**serious eye irritant, skin irritant, harmful if swallowed**
TLC plates (plastic-backed and containing fluorescent indicator, cut to approx. 6.5 cm × 2.5 cm)	not a hazardous substance
ethyl acetate	**flammable, irritant to eyes**
hexane	**highly flammable, skin irritant, may cause drowsiness**

Equipment: TLC eluting tank – either a suitable-sized jar with lid or 100 cm³ beaker covered with a watch glass or aluminium foil, 10 cm³ measuring cylinder, small sample vials, thin capillary tubes, filter paper, UV 254/365 nm lamp

Procedure

1. Prepare the elution solvent of 5% ethyl acetate in hexane by mixing 0.5 cm³ of ethyl acetate and 9.5 cm³ of hexane in a 10 cm³ measuring cylinder.
2. Transfer a few milligrams of cholesterol (starting material) and the recrystallised cholesteryl benzoate to separate small sample vials. Dissolve the solids in a small amount of the elution solvent.
3. Without pressing hard enough to disturb the silica surface, draw a pencil line about 1 cm from the edge of a TLC plate.
4. Mark two pencil crosses, evenly spaced, along the pencil line on the TLC plate.
5. Place a thin capillary tube in each vial for a few seconds; some of the liquid will rise up into the capillary tube.
6. Gently apply the tip of the capillary containing cholesterol to the left-hand cross. The liquid should flow out of the end of the capillary and spread out on to the cross. Remove the capillary from the TLC plate once the spot is about 1–2 mm in diameter.
7. Spot the cholesteryl benzoate solution onto the right-hand cross in the same manner as cholesterol.
8. Transfer approx. 5 cm³ of the elution solvent (prepared in step 1) to the eluting tank. *The liquid must not come high enough for the material spotted onto the crosses to be covered by solvent.*
9. Stand a piece of filter paper (which has been cut to give a flat edge a few cm in length) inside the eluting tank. Wet the filter paper with the eluting solvent which helps in saturating the atmosphere in the eluting tank with solvent vapour.
10. Using a pair of tweezers to hold the top of the TLC plate (the opposite end from the pencil line), stand the plate in the eluting tank. Immediately cover the top of the eluting tank.
11. Leave the TLC plate in the tank until the solvent front has risen to about 5 mm from the top of the TLC plate.

12. Remove the TLC plate from the eluting tank and immediately mark the solvent front with a pencil.
13. Allow the TLC plate to dry.
14. Look at the TLC plate under the UV light (254 nm setting). Cholesteryl benzoate will show up as a dark spot on a green background. Draw around the spot with a pencil. Cholesterol is not visible under UV light as it does not have a chromophore which can absorb UV light.

    CARE *UV light is particularly hazardous and can cause serious damage to eyes and skin. Never look directly at the light source and it is recommended that gloves are worn. Ideally, the compound should be observed in a UV viewing cabinet.*
15. Place the TLC plate in a jar containing a few crystals of iodine. Iodine has a high vapour pressure for a solid and the jar will be saturated with iodine vapour. Leave the plate in the tank until it develops a light brown colour. Iodine has a high affinity for both unsaturated and aromatic compounds and both cholesterol and cholesteryl benzoate will appear as dark brown spots on a lighter brown background.

    CARE *Carry out this step in a fume cupboard.*
16. Circle the cholesterol and cholesteryl benzoate spots with a pencil. If the reaction has gone to completion there should be no cholesterol in the sample of cholesteryl benzoate.
17. The relative movement of an individual compound up a TLC plate is expressed in terms of its R_f (retention factor) value. Using a ruler, measure the distance from the baseline to the centre of the spots and also the distance between the baseline and solvent front. The R_f is the ratio of the two values, i.e. R_f = distance moved by the compound spot ÷ distance moved by the solvent.

    The R_f values give information on the relative polarity of the compounds; non-polar compounds move a long way up the plate whilst polar compounds move very little.

6.6.2.3   Spectroscopic Evaluation

IR and NMR spectroscopic techniques are based on the interaction of electromagnetic radiation with molecules. By analysing spectra obtained from these and other techniques there is often sufficient information to identify the structure of a molecule. IR spectroscopy is a means of identifying the type of covalent bonds and functional groups in a molecule whilst ^{1}H NMR spectroscopy provides information about the hydrogen environments in a molecule. The theory underlying IR and NMR spectroscopy and how the spectra are interpreted are complex and only a very brief outline with salient points is given below. Readers interested in these areas are directed to some of the specialist books in the *Further Reading* section at the end of the chapter. Whilst interpreting IR and NMR spectra is a complex process, there are some straightforward deductions that can be made about the progress of the reaction by comparing the spectra of the starting material (cholesterol) with that of the product (cholesteryl benzoate).

6.6.2.3.1 IR Spectra

Bonds connecting atoms in molecules stretch and bend, i.e. they vibrate, and a useful parallel is a spring connecting two masses. The frequency, and therefore the energy, of vibration of the spring depends on the strength of the spring and on the masses present. In molecular systems, the frequency of vibration is a function of the masses of the atoms and the strength of the bond. For example, the energy required to vibrate an O–H bond will differ from that required to vibrate a C=O bond. Vibrating a bond uses energy that is found in the IR region of the electromagnetic spectrum. When IR radiation of the same frequency as that of a particular vibrating bond interacts with the bond then it is absorbed, and this change is detected in the IR spectrum. An IR spectrum is a plot of wavenumber, the reciprocal of the radiation wavelength as the x-axis (with units of cm^{-1}), against transmission of IR radiation as the y-axis. The usual range on the x-axis is between 4,000 and 400 cm^{-1} with absorption peaks indicative of functional groups occurring in the left-hand part of the spectrum in the region between 1,500 and 4,000 cm^{-1}. The region to the right of 1,500 cm^{-1} is known as the fingerprint region and is complex and difficult to interpret. Whilst some absorptions assignable to functional groups occur in this region there are many that correspond to vibrations of the molecule as a whole. As the latter peaks are unique to a particular molecule the fingerprint region is useful to confirm the identity of a molecule by comparison with authentic spectra in a database.

In the IR spectrum of cholesterol (Figure 6.9) there are few significant peaks above 1,500 cm^{-1}: one at 3,412 and several clustered around 3,000 cm^{-1}. These latter peaks are the stretching vibrations of C–H bonds and absorptions in this region are seen in most organic compounds. The diagnostic absorption is that at 3,412 cm^{-1} which is due to an O–H group and hydrogen bonding effects result in a broad peak. The absence of an O–H peak in the cholesteryl benzoate spectrum (Figure 6.10) is informative as it indicates that the hydroxyl group has reacted. The reaction to form cholesteryl benzoate creates an ester group (-CO–O-) which features a C=O and a C–O bond. The C=O bond is very distinctive in the IR spectrum and the strong sharp peak at 1,709 cm^{-1} in the cholesteryl benzoate spectrum is a typical value, falling in the middle of the full C=O range between 1,850 and 1,650 cm^{-1}. The C–O stretching vibration occurs in the fingerprint region, but as cholesterol and cholesteryl benzoate both feature a C–O bond, identifying these absorptions is not definitive. The two small peaks at 1,601 and 1,580 cm^{-1} in the cholesteryl benzoate spectrum are aryl carbon-carbon bond vibrations and indicate the presence of an aromatic ring in the compound.

Inspection and comparison of the IR spectra of cholesterol and cholesteryl benzoate suggest that the esterification reaction carried out was successful and this conclusion is also supported by evidence from the NMR spectra.

6.6.2.3.2 NMR Spectra

Atomic nuclei have a property called spin which causes some nuclei to behave as small magnets. Hydrogen nuclei (or protons), for example, have two possible orientations when placed in a magnetic field, either aligned with the field or against the field. The two orientations have different energies with the orientation aligned with the

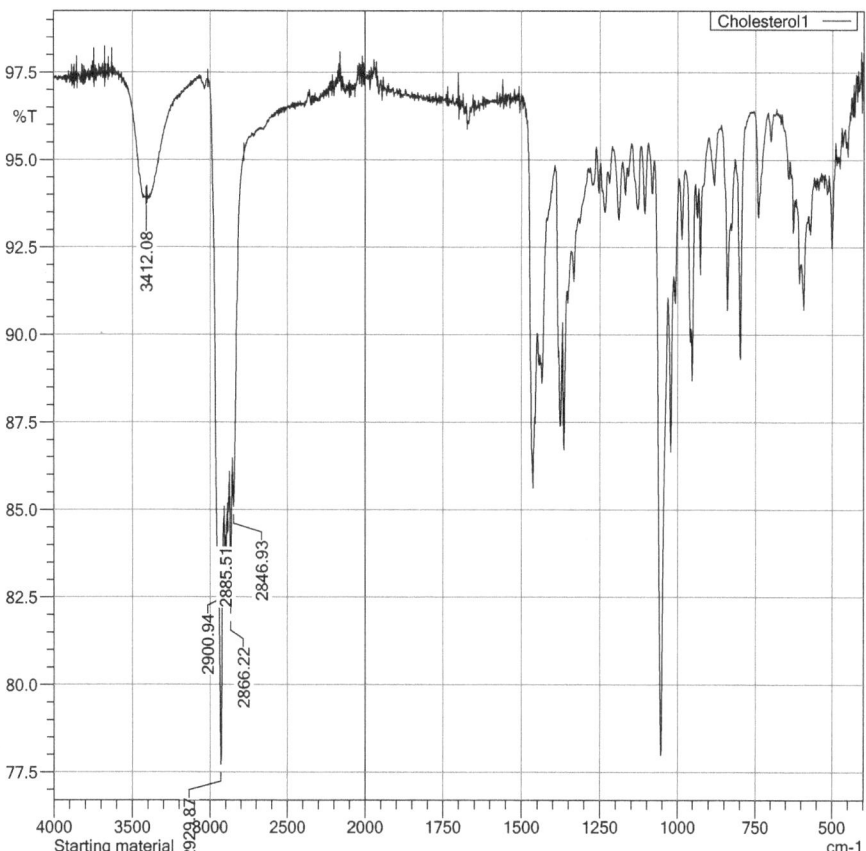

FIGURE 6.9 Infrared spectrum of cholesterol. The sample was run as the solid crystalline material on a Shimadzu IRTracer-100 instrument.

field having a lower energy than that aligned against the field. Transitions between these two states occur through the absorption of energy which corresponds to the radio wave region of the electromagnetic spectrum. When the energy gap between the two states matches the frequency of the applied radiation it is known as resonance. Electrons within a molecule shield the nuclei to some extent from the external magnetic field and this causes different atoms in the molecule to absorb slightly different frequencies of radiation. An NMR spectrum is a plot of chemical shift (δ), a measure of the nuclear shielding effect, as the x-axis against absorption of energy as the y-axis. Chemical shift is expressed as parts per million (ppm) with respect to an internal reference compound or signal and in proton NMR spectroscopy the chemical shift scale covers the range $\delta = 0$–10 ppm.

Cholesterol and cholesteryl benzoate have 46 and 50 hydrogen atoms respectively in their structure. Many of these hydrogen atoms are in very similar environments, i.e. as methyl (CH_3), methylene (CH_2) or methine (CH) groups in the molecules and give rise to

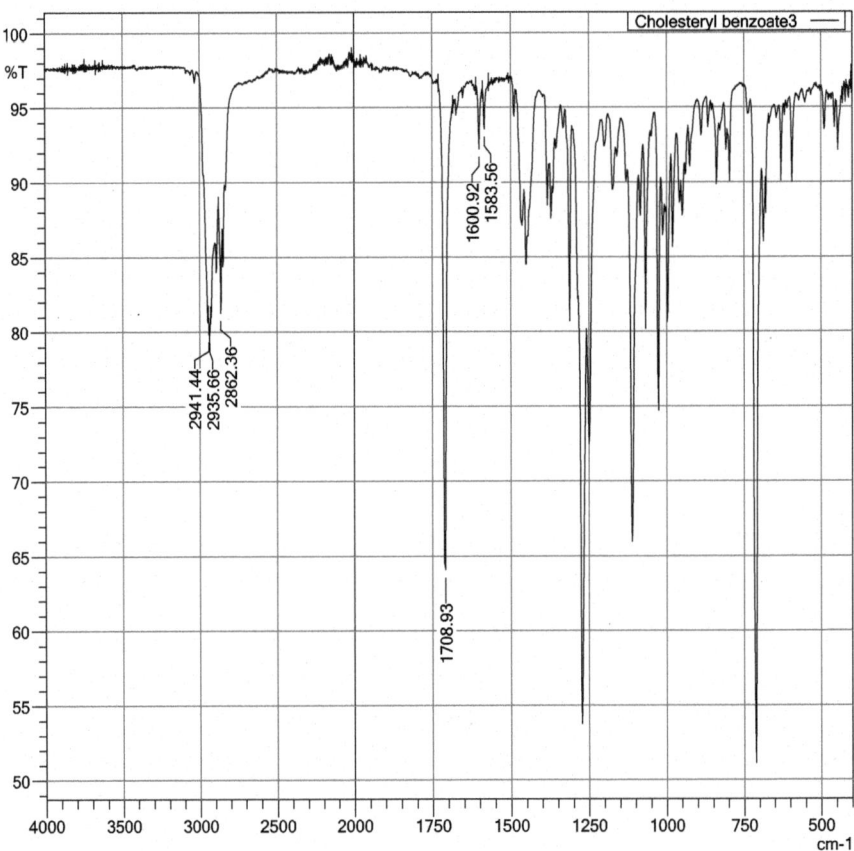

FIGURE 6.10 Infrared spectrum of cholesteryl benzoate. The sample was run as the solid crystalline material on a Shimadzu IRTracer-100 instrument.

peaks in the 0–2.5 ppm region of the ^{1}H NMR spectra (Figures 6.11 and 6.12). An electro-negative atom such as oxygen attached to the carbon of a C–H bond results in the chemical shift of the proton to move to a higher δ value (known as downfield). Alkene and aromatic protons are also shifted downfield with C=C–H protons typically being found between 4.5 and 6.5 ppm and Ar–H between 7 and 9 ppm. The assignments for the downfield shifted protons in cholesterol and cholesteryl benzoate are shown in Figure 6.13. The integration trace on each spectrum shows the relative size of the peak area (seen as a series of stepped lines above the peaks with numerical value below the x-axis) and indicates how many protons have the δ value shown. The chemical shift of a proton in a hydroxyl group (OH) can vary due to hydrogen bonding effects and calculations involving the integration values suggest that the OH peak occurs in the 0–2.5 ppm region.

The presence of the aromatic ring in cholesteryl benzoate shifts the position of the C-3 proton downfield compared to the δ value of this proton in cholesterol. This effect of the aromatic ring also influences the δ position of other nearby protons

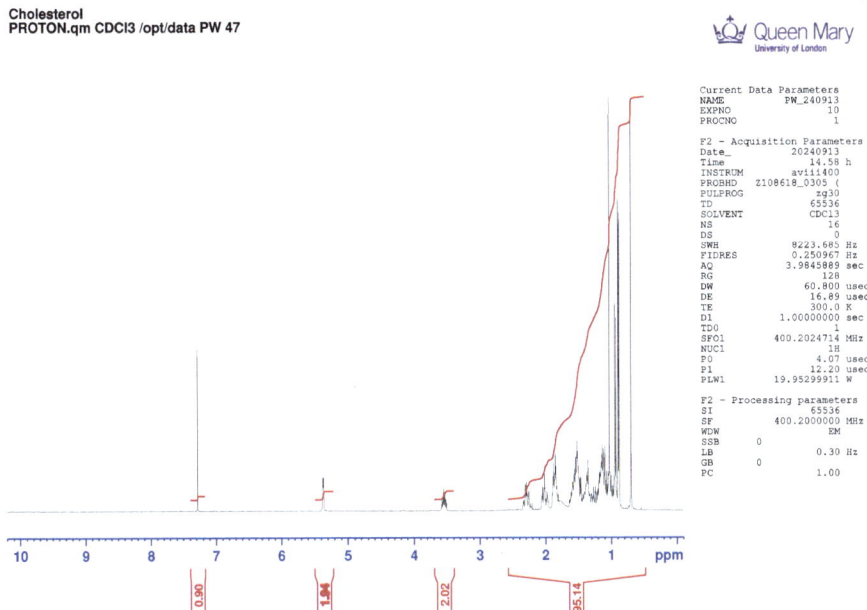

FIGURE 6.11 ¹H NMR spectrum of cholesterol dissolved in CDCl₃ solution. The sample was run on a AVIII400 MHz Bruker instrument. The peak at 7.2 ppm is due to the CHCl₃ impurity which is always present in CDCl₃ solvent.

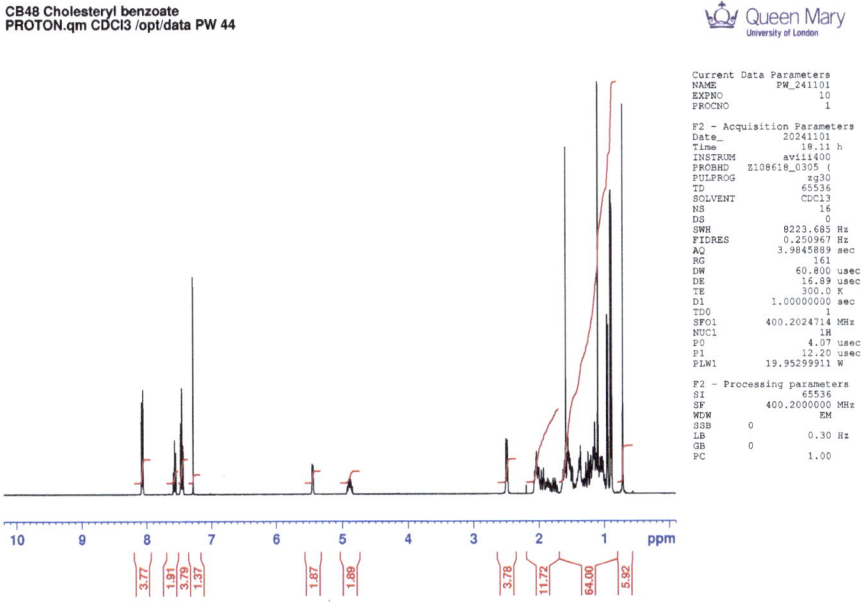

FIGURE 6.12 The ¹H NMR spectrum of recrystallised cholesteryl benzoate dissolved in CDCl₃. The sample was run on a AVIII400 MHz Bruker instrument. The peak at 7.2 ppm is due to the CHCl₃ impurity which is always present in CDCl₃ solvent.

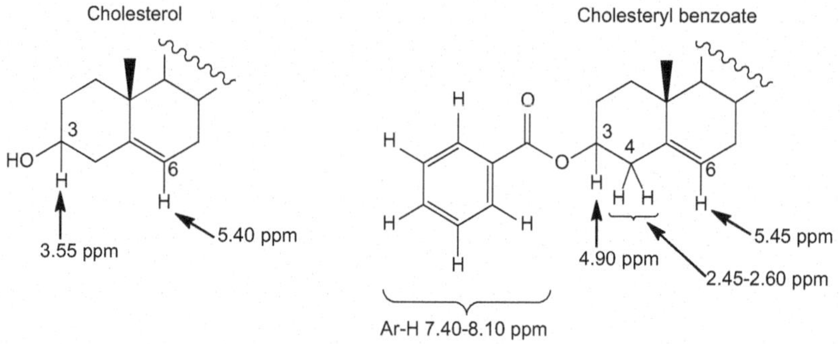

FIGURE 6.13 Assignments of peaks between 3 and 8 ppm in ^{1}H NMR spectra of cholesterol and cholesteryl benzoate. Note that the peak at 7.2 ppm is ignored as it is due to an impurity (CHCl$_3$) in the CDCl$_3$ solvent.

such as those on C-4, but to a lesser degree. The CH$_2$ protons at C-4 in cholesteryl benzoate resonate at 2.45–2.60 ppm and are distinctly separate from the crowded spectral region between 0.5 and 2.0 ppm. The resonances of the protons at carbons 3, 4 and 6 and the aromatic protons have fine structure which is due to spin-spin coupling. Interpretation of these patterns can give other useful structural information but this area is not considered in this very brief overview of ^{1}H NMR spectroscopy.

NOTE

1   The formula of ammonium iron(III) citrate is non-stoichiometric and is a mixture with an indefinite composition containing Fe^{3+} and NH$_4^+$ cations and citrate anions.

FURTHER READING

Additional details on the well-established experimental methods described in this chapter can be found in a range of books, such as those listed below. Prussian blue analogues, metal ligand complexes, liquid crystals and nanoparticles are all active research fields and the academic articles listed give insight into some of the experimental work. Other references on liquid crystals can be found in the *Further Reading* section of Chapter 5.

BOOKS

Burns, R. (2023) *Dyeing Yarn Naturally*. Ramsbury: Crowood Press.
Chalmers, A. (2024) *Creative Cyanotype: Techniques and Inspiration*. Ramsbury: Crowood Press.
Dean, J. R., Jones, A. M., Holmes, D., Reed, R., Jones, A. and Weyers, J. (2011) *Practical Skills in Chemistry*. 2nd Ed. Harlow: Prentice Hall.
Williams, D. H. and Fleming, I. (2007) *Spectroscopic Methods in Organic Chemistry*. 6th Ed. London: McGraw-Hill.

ARTICLES

Hurlbutt, K., Wheeler, S., Capone, I. and Pasta, M. (2018) Prussian Blue Analogs as Battery Materials. *Joule*. [Online] 2 (10). pp. 1950–1960. Available from: https://doi.org/10.1016/j.joule.2018.07.017. [Accessed: 19th November 2024].

Ielo, I., Rando, G., Giacobello, F., Sfameni, S., Castellano, A., Galletta, M., Drommi, D., Rosace, G. and Plutino, M. R. (2021) Synthesis, Chemical-Physical Characterization, and Biomedical Applications of Functional Gold Nanoparticles: A Review. *Molecules*. [Online] 26. Article no. 5823. Available from: https://doi.org/10.3390/molecules26195823. [Accessed: 19th August 2024].

Li, Z., Dellali, A., Malik, J., Motevalli, M., Nix, R. M., Olukoya, T., Peng, Y., Ye, H., Gillin, W. P., Hernández, I. and Wyatt, P. B. (2013) Luminescent Zinc(II) Complexes of Fluorinated Benzothiazol-2-yl Substituted Phenoxide and Enolate Ligands. *Inorganic Chemistry*. [Online] 52. pp. 1379–1387. Available from: https://doi.org/10.1021/ic302063u. [Accessed: 30th March 2025].

Lin, T.-H., Guo, D.-Y., Chen, C.-W., Feng, T.-M., Zeng, W.-X., Chen, P.-C., Wu, L.-Y., Guo, W.-M., Chang, L.-M., Jau, H.-C., Wang, C.-T., Bunning, T. J. and Khoo, I. C. (2024) Directed Crystalline Symmetry Transformation of Blue-Phase Liquid Crystals by Reverse Electrostriction. *Nature Communications*. [Online] 15. Article no. 7038. Available from: https://doi.org/10.1038/s41467-024-51408-4. [Accessed: 15th November 2024].

Sivakavinesan, M., Vanaja, M. and Annadurai, G. (2021) Dyeing of Cotton Fabric Materials with Biogenic Gold Nanoparticles. *Scientific Reports*. [Online] 11. Article no. 13249. Available from: https://doi.org/10.1038/s41598-021-92662-6. [Accessed: 28th May 2024].

Appendix

TABLE A.1
Mohs Scale of Hardness

Scale	Mineral	Simple Test Using Common Object
1.	Talc or graphite	Feels soapy or greasy
2.	Gypsum	Finger nail 2.5
3.	Calcite	
4.	Fluorite	Copper coin
5.	Apatite	5.5–6 window glass
6.	Orthoclase	6.5 steel blade or nail
7.	Quartz	
8.	Topaz	8.5 masonry drill-bit
9.	Corundum (ruby or sapphire)	
10.	Diamond	

Mohs scale is a comparative scale and is not linear. For example, diamond is roughly four times harder than corundum which is twice as hard as topaz. A soft mineral is one which has a hardness value of 1–2, a mineral with a medium level of harness has a value of 3–5 and a value of 6–10 indicates a hard mineral, one which can scratch glass but which is not scratched by a steel penknife.

NAME of ROCK ....

	IGNEOUS		SEDIMENTARY		METAMORPHIC

Main rock

Colour	1 (light)	2	3	4	5 (dark)
Lustre	1 (shiny)	2	3	4	5 (dull)
Texture	1 (smooth)	2	3	4	5 (rough)
Particle size	Fine (< 1 mm)		Medium (1-5 mm)		Coarse (> 5 mm)

Crystals within rock (there may be more than one type)

Colour	1 (light)	2	3	4	5 (dark)
Percentage	Low (< 20%)		Medium (~ 50%)		High (> 80%)
Arrangement of crystals	Scattered (smooth)				Layered (rough)
Crystal size	Fine (< 1 mm)		Medium (1-5 mm)		Coarse (> 5 mm)
Shape	1 (angular)	2	3	4	5 (round)

Mineral group(s)

Carbonates   Oxides   Silicates   Sulfides   Sulfates   Phosphates   Other

Formula of mineral(s):

Any other distinguishing feature(s) (e.g. variety of colours, if particularly lightweight or heavy, layered structure, wavy bands of dark and light minerals, mineral hardness)	Sketch

FIGURE A.1 An example of a chart to record key features of the bulk physical properties of rock samples upon initial examination. The mineral group reflects the chemical composition of the crystalline components present within the rock and more than one type is often present. The mineral composition is represented by the chemical formula and the specific group is identified by the negative or anionic part of the formula. When samples are viewed in the field the location and/or map coordinates should also be recorded.

NAME of FOSSIL...

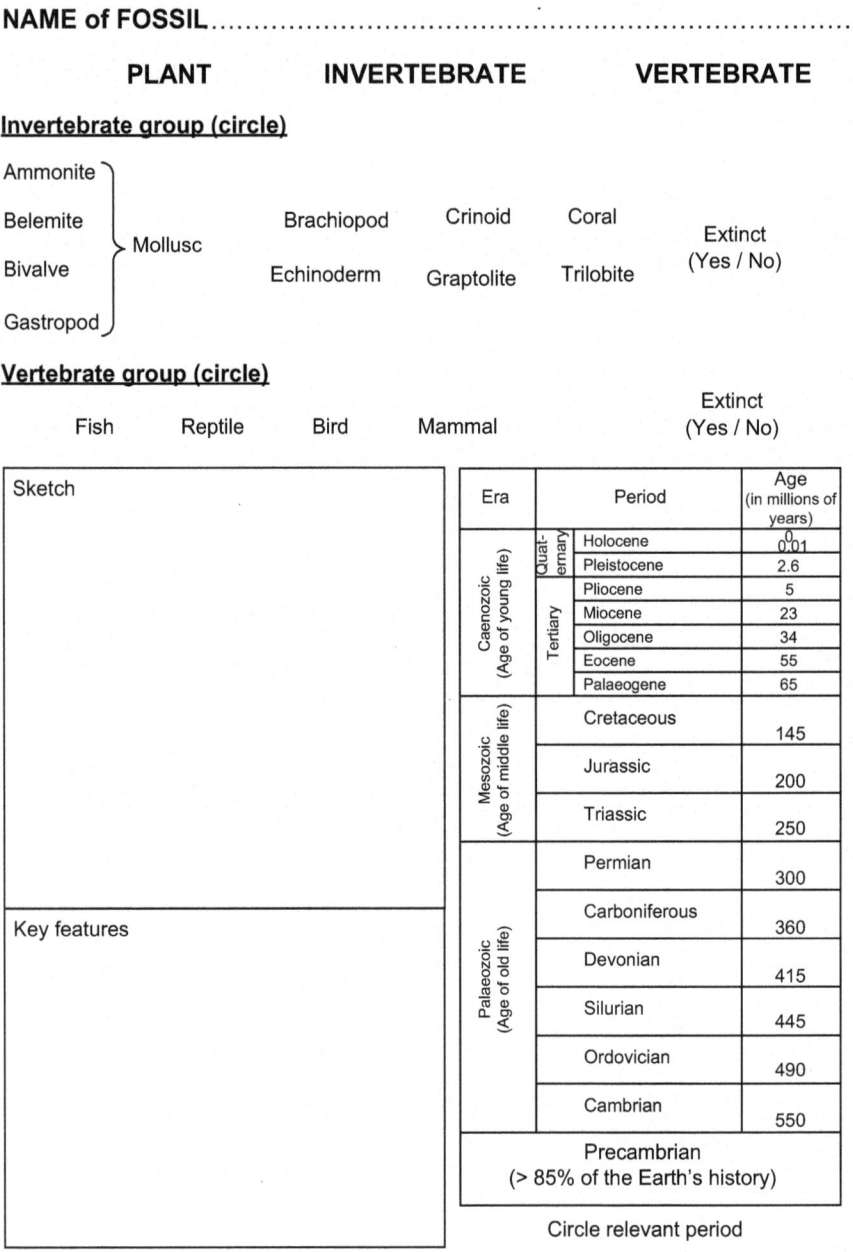

FIGURE A.2   An example of a chart to record key features of fossil samples upon initial examination.

Glossary

Agonist: A drug activating a receptor.

Alchemy: The forerunner of modern chemistry; concerned with the transmutation of metals (in particular with attempts to convert base metals into gold), to find a universal cure for diseases and an elixir of immortality.

Alloy: A material consisting of two or more metals, e.g. bronze is an alloy of copper and tin.

Analogue: (medicinal chemistry) A compound with a molecular structure that is similar to another but which differs from it in certain components such as functional groups or substructures.

Analogue: (photography) The practice of using a film camera to capture an image and the application of chemical processes to produce a photograph on film or paper.

Angiogenesis: Formation of new blood vessels; an anti-angiogenesis effect is when a drug prevents new blood vessels from forming.

Anisotropic: Refers to when the physical properties of a material such as refractive index or electrical conductivity vary depending on the direction of measurement.

Antagonist: A drug which interacts with a defined receptor to block an agonist.

Anticline: An upfold in the form of an arch.

Antigen: A molecular species which can stimulate an immune response in the body.

Artefact: (archaeology) An item made or shaped by humans such as a tool or work of art.

Bedrock: The continuous solid rock of the continental crust.

Biologics: Very large and complex molecules produced using recombinant DNA technology and used as drugs, e.g. monoclonal antibodies.

Biostratigraphy: The part of stratigraphy that utilises fossils to date rocks.

Biosynthesis: The production of molecules by a living cell.

Birefringence: (double refraction) The property exhibited by certain molecularly ordered materials of forming two refracted rays from a single incident ray.

Calcination: The heating of an ore in order to decompose it and to drive off volatile products.

Carcinogenic: Having the potential to cause cancer.

Cellulose: A polymer of glucose, contained in paper and cotton, with many thousands of repeat units in a chain.

Chromophore: The structural unit in a molecule responsible for the absorption of radiation. Absorption of light in the visible region of the spectrum results in a coloured compound.

Cleavage: (in minerals) The tendency of a mineral to break in preferred directions along the plane surfaces.

Cryo-EM: (cryogenic electron microscopy) Structure determination of delicate biological molecules, such as membrane proteins. The procedure involves flash freezing the sample at extremely low temperatures, which preserves its native three-dimensional structure; the frozen material is then imaged using an electron microscope.

Diurnal: An animal or plant that is active during the day.

Dolmen: A prehistoric monument consisting of two or more upright stones with a large flat stone lying across them; they do not have the shaped and jointed stones found in the trilithon structures of Stonehenge.

Drug: A chemical substance that acts on biological systems or functions to treat, prevent or alleviate disease and illness.

Enantiomers: Stereoisomers related to each other as non-superimposable mirror images.

Endogenous: Produced within an organism as a result of normal physiological processes.

Fast: (textiles) How resistant a dye or a dyed material is to losing its colour under certain conditions; e.g. when exposed to sunlight (lightfast) or during normal washing (colourfast).

Fault: Fracture along which opposite sides have been displaced relative to each other.

Fossil: The preserved remains or traces of an ancient plant or animal.

Gangue: The nonvaluable minerals of an ore.

Genomics: Identification and characterisation of all the DNA (i.e. the genome) of an organism to understand how genes function and interact with each other.

Glycoside: A molecule with an attached sugar unit such as glucose that on hydrolysis gives the sugar and the nonsugar component; the molecule without the sugar is called an aglycone.

Half-life: (in radioactive decay) The time required to reduce the amount of the parent element by half.

Hardness: (mineral) The relative resistance of a mineral to scratching.

Hormone: A molecule secreted by glands and organs of the endocrine system and transported through the bloodstream to other tissues and organs where it elicits a physiological response.

Hydrophilic: 'Water-loving'; a physical property of being attracted to water through the formation of hydrogen bonds, a property of polar molecules.

Hydrophobic: 'Water-hating'; a physical property of nonpolar molecules, *see* lipophilic.

In situ: A procedure carried out in the intended environment (from Latin for 'in the original place').

In vitro: Procedures using biological tissues, cells or molecules performed in a laboratory outside of a living organism (Latin for 'in glass').

In vivo: Biological processes occurring within living organisms or cells (Latin for 'within the living organism').

Intercalate: A process whereby a guest species (atom, ion or molecule) is trapped between layers in a host structure. There is no formal bonding between the host and the trapped species and the structural integrity of the host is maintained in reversible intercalation-extraction processes.

Interstitial site: The empty space or hole that exists between atoms in a close-packed lattice structure.

Ion channels: Macromolecules, often proteins, that span a lipid bilayer through which ions such as Na^+, K^+, Ca^{2+} and Cl^- can move from one side of the membrane to the other.

Isotropic: Refers to when the physical properties of a material are the same in all directions.

Lead compound: (medicinal chemistry) A compound showing a desired pharmacological property from which synthetic analogues are developed. Here lead rhymes with seed. Not to be confused with compounds containing the element lead (Pb).

Ligand: (inorganic chemistry) A molecule or ion that donates a pair of electrons to a metal atom or ion in forming a coordination complex.

Lintel: A horizontal piece of stone or wood above an opening in a building such as a door or window; can be decorative as well as providing structural support.

Lipophilic: A physical property of having an affinity with non-polar environments such as fats and lipids; synonymous with hydrophobic.

Mordant: A substance to allow the attachment of a dye to a fibre.

Morphology: A particular form or shape (of a crystal).

Non-stoichiometric: The elemental composition of a compound which cannot be represented by a simple integer ratio.

Olfactory: Sense of smell.

Orbital: A region of space in an atom or molecule within which there is a high probability of finding an electron.

Orogenesis: Process of mountain building; orogeny is the period of mountain building.

Oxidation state/number: A reflection of the positive or negative character of an atom in a compound; a number is assigned to the element in a molecule or ion on the basis of a set of formal rules. Roman numerals are used in systematic nomenclature.

Peptide: An organic compound composed of two or more amino acids and designated as di-, tri-, poly- etc. according to the number of amino acids linked by peptide bonds (CO–NH).

Petrography: The systematic description of rocks using hand specimens and thin sections.

Pharmacodynamic: The interaction of a drug with cells, the biochemical and biophysiological effects of drugs and the mechanism of their actions, i.e. what a drug does to the body.

Pharmacokinetic: The movement of a drug within the body encompassing its absorption, distribution, metabolism and elimination, i.e. what the body does to a drug.

Pharmacophore: The group of atoms and functional groups in a molecule and their relative positions in space which is responsible for its biological activity.

Pheromones: Molecules produced and released into the environment by one individual of a species (especially insects) which affects the sexual or social behaviour of other members of the same species.

Photon: A quantum (discrete unit) of light energy.

Physiology: The normal functions and processes of living organisms or their parts, such as organs, tissues or cells.

Polymorphs: Two or more distinct crystal forms of a single chemical substance.

Prodrug: A biologically inactive compound that is metabolised in the body to yield the active compound that is responsible for a drug's action.

Protein: A complex organic compound composed of hundreds or thousands of amino acids.

Proteomics: Complete evaluation of the function and structure of proteins in order to understand how they interact with one another inside cells.

Racemic: Containing a 1:1 mixture of two enantiomers.

Radiogenic: Produced by radioactive decay.

Receptor: A macromolecule, typically a protein, that binds to a specific substance and causes a specific effect in a cell.

Recombinant DNA: (rDNA) A DNA molecule created by ligating segments of DNA that normally are not contiguous.

Refractive index: A measure of the extent to which a medium refracts light.

Roasting: (metallurgy) Heating an ore in air to form the metal oxide prior to smelting; it is particularly important in processing sulphide ores.

Smelting: Heating a metal ore to a high temperature, often in the presence of a reducing agent, to cause a reaction in which the metal product is in a liquified state and this aids its recovery.

Stalagmite: A blunt, icicle-like column of rock projecting upwards from the floor of a cave.

Stereoisomers: Molecules in which the atoms are joined together in the same order but have a different orientation in space. Optical isomers are asymmetric molecules due (usually) to the presence of one or more chiral centres; a tetrahedral carbon atom with four different groups attached. Geometric or E-Z isomers occur when there is restricted rotation in a molecule, most commonly due to the presence of a double bond. Prefixes E and Z (sometimes *cis* and *trans*) are used to indicate the different spatial arrangements of atoms or groups in geometric isomers. In optical isomerism the prefixes R and S are used.

Stratigraphy: The branch of geology dealing with rock successions and geological history.

Structure-activity relationships: Studies carried out to determine those atoms or functional groups which are important for a drug's activity. Quantitative structure-activity relationships (QSAR) use mathematical models to predict various drug-related properties.

Syncline: A downfold with a troughlike form.

Teratogen: A substance that interferes with normal embryo or foetal development and causes congenital malformations.

Trilithon: A structure consisting of two large vertical stones and a third stone anchored horizontally across the top as a lintel.

Index

Note: **Bold** page numbers refer to tables, *italic* page numbers refer to figures and page numbers followed by "n" refer to end notes.